AEROSPACE EXPLORATION

航天探秘

在航天场景中学英语

主编　肖志军

中国出版集团

研究出版社

图书在版编目（CIP）数据

航天探秘——在航天场景中学英语 / 肖志军主编 . –– 北京 : 研究出版社 , 2021.1

　ISBN 978–7–5199–0932–1

　Ⅰ . ①实… Ⅱ . ①肖… Ⅲ . ①航天学—英语 Ⅳ . ① V4

中国版本图书馆 CIP 数据核字 (2020) 第 193850 号

出 品 人：赵卜慧
图书策划：张　琨
责任编辑：张　璐　张　琨

航天探秘——在航天场景中学英语
AEROSPACE EXPLORATION:LEARNING ENGLISH IN AEROSPACE
SCENARIOS

主　编	肖志军	
编　委	肖志军　张园园　吴　昊　贺　利	
	张玉梅　周万隆　陈凌晖	
出版发行	研究出版社	
地　址	北京市朝阳区安定门外安华里 504 号 A 座（100011）	
电　话	010–64217619　　64217612（发行中心）	
网　址	www.yanjiuchubanshe.com	
经　销	新华书店	
印　刷	河北赛文印刷有限公司	
版　次	2021 年 1 月第 1 版　2021 年 1 月第 1 次印刷	
开　本	710 毫米 ×1000 毫米　1/16	
印　张	17	
字　数	180 千字	
书　号	ISBN 978–7–5199–0932–1	
定　价	66.00 元	

前　言

　　《航天探秘——在航天场景中学英语》这本书的选题起源于一次出差闲聊。2019 年 8 月我和研究出版社的张琨老师去上海出差，参加我主编的《航天少年——学生空间搭载实验》三册图书的发布活动，张琨老师也是这套图书的责任编辑。聊天过程中，她得知我除了担任《航天员》杂志的执行主编外，还长期从事航天国际交流和航天员专业英语教学工作，多次承担载人航天新闻发布会的现场翻译任务，当即提出能否撰写一本航天英语口语方面的图书。她提出市场上并没有类似的图书，这本书如果能出版，将填补"航天英语"这个市场空白。

　　坦白说，我也曾经萌生过撰写航天专业英语图书的念头。在多年从事载人航天国际交流工作中，也有意识地积累了大量资料和素材，但由于一直没有想好切入点，写书的事就一直搁置下来。张老师的选题意识非常敏锐，她的建议与我的想法不谋而合。于是，我爽快地答应下来，并初步约定 2020 年上半年交稿。

　　为了避免专业英语的枯燥乏味，这本书围绕着"场景"二字大做文章。我在编写过程中精心设计了人物和故事情节，将全书各章有机结合起来。凯文是一位资深航天发烧友，也是中国航天员的铁杆粉丝。刚刚取得某大学航天专业的录取通知书的凯文，幸运地赢得一次参观北京航天城的机会，并与航天专家进行了深度对话，本书的主题内容也随着场景的变化，逐渐展开。

本书以对话的形式系统介绍了航天的各个领域，覆盖面广，内容新颖精炼。全书共分六章，涉及航天中心与设施、如何成为航天员、进入太空、太空生活、无人探索和未来计划等主题。为了便于读者学习，每章除对话外，还对精心挑选的专业词汇及缩略语进行了注解，并补充5篇专业的扩展阅读材料。在本书的编辑过程中，尤其重视图书的"实用性"，无论是情景对话、精选词汇、还是拓展阅读，都主要从航天员训练和真实工作环境中选取素材。希望《航天探秘——在航天场景中学英语》既可以成为航天领域相关人员、航天英语爱好者的案头学习资料，更可以成为航天英语教学的专业教材。

在本书的提纲策划和样章撰写过程中，张琨老师给予了很多指导和帮助，提出了非常专业的意见和建议。本书第二章第2节初稿由吴昊撰写，第四章第1-3节初稿由张圆圆撰写，贺利、张玉梅、周万隆、陈凌晖等分别参加了中文说明撰写、阅读素材搜集和词汇注解等工作。本书写作出版过程中，也得到了中国航天员中心机关和领导的大力支持。

衷心感谢张琨老师等各位参与者的辛勤劳动！感谢航天员中心的大力支持！感谢研究出版社的厚爱！

肖志军

CONTENTS

Chapter 1
Space Centers and Facilities

凯文是一位刚刚参加完高考的学生，顺利取得了某大学航天专业的录取通知书。凯文也是一位资深航天发烧友、中国航天员的铁杆粉丝。机缘巧合，暑假期间他获得了一次参观北京航天城的机会。

凯文参观北京航天城时，与航天专家进行了深度对话，系统了解了中国航天员科研训练中心和北京飞行控制中心承担的中国载人航天任务、科学研究以及设备设施。

他们还谈论了世界上著名的航天发射场，包括中国的酒泉卫星发射中心、海南发射场、位于哈萨克斯坦的俄罗斯拜科努尔航天发射场、美国肯尼迪航天中心、位于法属圭亚那的欧洲库鲁航天发射场；也谈论了国外的航天员中心，包括位于莫斯科星城的俄罗斯加加林宇航员训练中心、位于休斯顿的美国约翰逊航天中心、以及位于德国科隆的欧洲航天员中心。本章系统介绍了世界著名航天发射场和航天员中心。

Section 1. Touring Beijing Space City

场景介绍：北京航天城是中国航天员的摇篮，中国航天员中心和北京飞行控制中心坐落其中，凯文与航天专家现场交流，了解航天员的典型训练设备和设施……

Kevin has passed the college entrance examination this summer and is admitted to Beihang University. This summer vacation will be the most relaxing time for him in his entire life. Kevin is a fan of Yang Liwei, the first astronaut of China. He has read many stories about the Chinese astronauts and his dream is to be an astronaut in the future. To his excitement, he luckily wins an opportunity to visit Beijing Space City in a weekend. Mr. Ivan, a professor of Beijing Space City, explains the details of different facilities.

Astronaut Center of China

Ivan: Welcome to Beijing Space City! I am Ivan and I work in Astronaut Center of China, ACC.

Kevin: Thank you, Mr. Ivan! My name is Kevin. I am so excited about the visit!

Ivan: The first place we are going to visit is ACC. The full name of ACC is China Astronaut Research and Training Center. You know, this name is a little bit long, so we often use the name Astronaut Center of China instead.

Kevin: The main job of ACC is astronaut selection and training, is that right?

Ivan: Yes, astronaut selection and training is the most important work of ACC. You will see some of the training facilities during the visit. Nevertheless, ACC also has many other responsibilities.

Kevin: Can you explain?

Ivan: Ok. In addition to the astronaut selection and training, ACC also studies and manufactures many space products such as the space suits, the space food, the environment control and life support system, the space medical monitoring devices etc. Basic researches in space medicine and human factors engineering are also conducted in ACC.

Kevin: Can you say something about those basic researches?

Ivan: ACC has two national key laboratories. One is called the State Key Laboratory of Space Medicine Fundamentals and Application, which mainly studies the adverse effects of space environment on the human body and how to mitigate those side effects. The other is called the National Key Laboratory of Human Factors Engineering, which studies how the human and machine better match each other to improve the performance of a system. These studies can help ensuring the safety, health and good performance of astronauts.

Kevin: That sounds difficult!

Ivan: Maybe. Next, I will show you something not so difficult. Let's go to the astronaut training hall.

Kevin: Great!

Ivan: This is the mock-up of Tiangong-2. The inside and the size of this mock-up are exactly the same with Tiangong-2 flown in space. The facility beside Tiangong-2 mock-up is called Fix-based Spaceship Training Simulator.

Kevin: Can I see the inside of the simulator?

Ivan: Yes, you can. Just like the aviation simulator, this simulator is used by astronauts to learn spaceship operations. The upper part is the orbit module of the Shenzhou spaceship and the lower part is the return capsule. During training, the crew, wearing pressure suit, practice all kinds of operations of the spaceship. They often have to stay inside the simulator for several hours just to finish one training sortie.

Kevin: How does the astronaut go to the bathroom during training?

Ivan: They will not go to the bathroom. Instead, they use diapers, just like the real flight.

Kevin: Can this simulator simulate the micro-gravity environment?

Ivan: No, it cannot. It just simulates all kinds of flight procedures. The astronaut can be trained here to master the normal procedure, the abnormal procedure and the emergency procedure.

Kevin: What is the smaller facility on the right?

Ivan: That is also a spaceship simulator, but it is dedicated for rendezvous and docking training or RVD training. RVD means the meeting and connecting of two space vehicles. With RVD technology, we can transport crew and resupply cargos into space.

Kevin: Amazing!

Ivan: Let's visit another training site, the human centrifuge.

Kevin: I have seen it in a TV program, it rotates and creates higher gravity, right?

Ivan: Yes. Astronauts trains their hyper-gravity tolerance here on this human centrifuge. The centrifuge will create up to 8G during the training, that means the stress is eight times the weight of the body. It is very challenging! But if you want to be an astronaut, you must pass this test.

Kevin: It seems very hard, but I am not afraid!

Ivan: Let's visit the next site, the NBL.

Kevin: What is NBL?

Ivan: NBL is the abbreviation of Neutral Buoyancy Laboratory. It is also called the weightlessness simulation water tank. In colloquial language, it is often called the "pool".

Kevin: I can see some modules inside the pool.

Ivan: Yes, they are the mockups of the spacecraft and segments of the China space station.

Kevin: What is the purpose of this pool?

Ivan: It is mainly used for EVA training, or space walk training. EVA means extravehicular activity. Of course, it also serves as a research facility to do some experiments, for instance, verification of EVA procedures, or human factors researches.

Kevin: How are the astronauts trained here?

Ivan: I will show you a video, and you are able to see the complete EVA training process.

Kevin: That's amazing! Why do astronauts take this training?

Ivan: You know, astronauts in space will experience weightlessness or micro-gravity, so they must learn how to work in micro-gravity.

Kevin: How long does this training take?

Ivan: It takes several hours for the astronauts to complete one training session. But the scuba divers who support the training underwater will change shifts every two hours. It is quite demanding physically for the astronaut, because he or she must always fight the residual pressure in the training suit.

Kevin: Are the training suits the same with EVA space suit in space?

Ivan: No, it is not the same. You see there is an umbilical cord connected to the underwater training suit which provides the oxygen and ventilation. In space, the real EVA space suit has a life support system in the backpack providing oxygen and removing the carbon dioxide exhaled.

Kevin: Thank you for the explanation!

Ivan: You are welcome! And that's all for the visit of ACC. Next, we will visit BACC, Beijing Aerospace Control Center.

Beijing Aerospace Control Center (BACC)

Kevin: I have seen this building many times on TV.

Ivan: Yes, almost all the space missions in China are controlled in this building. Let's go inside. First, let's see a short video about BACC. Then, we will go inside to visit the control hall.

Kevin: That's great!

Ivan: That's all for the video. You know, BACC was founded in 1996. But it has witnessed many milestone space missions of China, including the first unmanned spaceship mission, the first manned missions, the Tiangong mission, the Chang'e lunar exploration missions, etc. Soon we will see the construction of the China Space Station, and even the mission to Mars.

Kevin: So, when the astronauts are flying in space, all the communications and operations are controlled here?

Ivan: Yes. For any space flight mission, we must have a Tracking, Telemetry & Command and Communication System or TT&C and C system. It is just like flying a kite, we need a tread to control it.

Kevin: Is "Yuanwang" tracking ship part of the system?

Ivan: Yes. To ensure the communication coverage rate, we have not only the "Yuanwang" tracking ship, but also many fixed stations on the ground around the globe and the "Tianlian" relay satellites in space.

Kevin: What does the communication coverage rate mean?

Ivan: It means, when circling around the earth, the percentage of communication time you can have with the ground. Of course, 100 percent means the communication is established with the ground at any time.

Kevin: I understand. Thank you!

Ivan: Let's go inside to the control hall.

Kevin: Ok!

Ivan: The control hall is a two-story tall building. The large display in the front can show photos, videos, animations, and various parameters of the spacecraft. The first floor is the working place, you can see rows of consoles. The second floor is for visitors. Guests can watch space operations on the second floor.

Kevin: Marvelous! Are the state leaders talk to the astronauts in this room?

Ivan: Yes, this is the phone they used to make the space call. That's all for the visit of BACC. You can take a photo here.

Kevin: Thank you!

Section 2. Launch Sites

场景介绍：发射场是载人航天的一个重要系统，航天专家带你"走进"酒泉卫星发射中心和海南卫星发射场，并带你了解美国肯尼迪航天中心、俄罗斯拜科努尔航天发射场和欧洲库鲁航天发射场。

Jiuquan Satellite Launch Center

Kevin: I know there are several launch centers in China and Jiuquna satellite launch center is the most famous one.

Ivan: Yes, Jiuquan launch center is very famous, the first satellite of Dongfanghong-1, the Shenzhou spaceships and Tiangong space labs were lifted off from this center. Do you know where is the center located?

Kevin: It is in Jiuqan, I think.

Ivan: No. Jiuquan is a city of Gansu province. But the center is in Ejina Banner of Inner Mongolia. It is so named because Jiuquan is the only significant city in its vicinity.

Kevin: That is interesting! Jiuquan launch center is not in Jiuquan!

Ivan: Now, Jiuquan launch center is a man-made oasis in the Gobi-desert. I really admire people who have constructed the center in such a harsh desert! The rainbow bridge is the gate of the center.

Kevin: What sites can I visit in the center.

Ivan: The first site you must visit is the assembly building of Shenzhou Spaceship and the launch vehicle. The gate of the assembly building is the biggest in Asia. Several kilometers away opposite the assembly building is the launch pad. The Chinese scientists adopted the unique "vertical assembly, vertical testing and vertical transport" strategy for the launch mission.

Kevin: That's very interesting. Where do the astronauts live when they are here in the center?

Ivan: Wentiange is the crew quarter in the launch center. Two weeks before launch, astronauts will enter into preflight quarantine in this place. All the medical support and training before launch will take place here. There is also a media hall for the preflight interview. The departure ceremony you have seen on TV just happened here.

Kevin: What about the mission control facility?

Ivan: There is also a mission control hall here in the center, it is mainly used during the lift-off phase. In Jiuquan center, there is also a landing site.

Kevin: I know the landing site of Shenzhou spaceship is in Siziwangqi which is far away from Jiuquan, is that right?

Ivan: Yes, in the past, all the Shenzhou spaceships landed in Siziwangqi in Inner Mongolia. In Jiuquan center, there is a backup landing site. But in the future, the spaceships will land here in Jiuquan. The landing site in the Gobi desert will be the primary landing site!

Kevin: Any more places worth visiting there?

Ivan: Yes. The historic launch pad of Dongfanghong-1 is worth visiting. It is about half an hour's drive from the Dongfeng Space City. You can see the launch facilities, including the launch pad, the service tower and the underground control room.

Kevin: Thank you for the introduction!

Ivan: You are welcome!

Hainan Satellite Launch Center

Kevin: Professor Ivan, as far as I know we have already had three launch centers in China, why do we build Hainan satellite launch center?

Ivan: Hainan satellite launch center is in Wenchang city of Hainan province. The location is one of the reasons. Because the earth is rotating, the closer we are to the equator, the more payload we can launch into orbit for the same rocket. That's why the European space port was built in French Guiana.

Kevin: Any other reasons?

Ivan: Yes, another reason is the transportation of rockets. To launch heavy payloads into space, we need bigger rockets. But that will cause problems in transportation, it is impossible to transport launch vehicles like CZ-5. The core stage is about 5 meters in diameter, the current railway and high way system in China do not support the transportation of such big rocket.

Kevin: I see.

Ivan: So, the manufacture of such heavy rocket is in Tianjin and then it will be transported by ship to Wenchang for launch.

Kevin: When was Hainan center put in to service?

Ivan: Hainan Launch Site project was initiated in Aug., 2007, started construction in May, 2010, and put into service in 2014. It is a world's first-class, modernized, ecology-friendly launch center.

Kevin: What kind of missions will be undertaken in Hainan center?

Ivan: The Tianzhou cargo ship, the core module and experimental modules of China space station will be launched here. So does the Mars mission and new manned spaceships.

Kevin: What facilities can I visit there?

Ivan: You can visit the assembly building and the launch pad.

Kevin: Thank you for the introduction.

Baikonur Cosmodrome

Kevin: Where is the Russia human launch site?

Ivan: At present, the Russian cosmonauts are all launched from Baikonur Cosmodrome, but this place is not in Russia.

Kevin: Then where is it?

Ivan: It is in Kazakhstan. It has been the launch site for the Soviet Union and Russian space missions since the beginning of space exploration in 1957.

Kevin: How do the Russians use the launch site since it is not in the territory of Russia?

Ivan: Actually, it is an enclave of Russian territory. The Kazakh and Russian governments work together on the maintenance and operations of Baikonur, with Russia paying at least $115 million annually to lease the land.

Kevin: How far is Baikonur to Moscow?

Ivan: It is about 2100 km away. The launch complex is on a desert steppe. The location was selected mainly for its advantage in radio communication as well as its remoteness.

Kevin: Was the first cosmonaut Gagarin launched from this place?

Ivan: Yes, on April 12, 1961, Yuri Gagarin went into space from Baikonur. The first satellite Sputnik-1 was also lifted off from this launch site in 1957.

Kevin: Are the resupply missions and crew rotation missions to the International Space Station launched here?

Ivan: Yes, all the crewed Soyuz spaceships are launched from here, so do the Progress cargo spaceships. Manned spaceship launches have continued here for nearly 5 decades, including different versions of spacecraft and cosmonauts and astronauts from different nations.

Kevin: I know that the American astronauts buy tickets from Russia to go to the ISS. How much do they pay for one seat on the Soyuz spaceship?

Ivan: The Roscosmos charged NASA $21.8 million per seat in 2008, and the price went up to $81 million per astronaut in 2018. The latest price is expected to go up to $91 million per seat this year.

Kevin: The ticket is really expensive!

Ivan: Americans are expecting to buy the tickets from their own private partners in the future. This month we have witnessed the first commercial crew flight of SpaceX.

Kevin: What sites can I visit here?

Ivan: There are several sites you should visit: Site 1 is the launch pad for Soyuz, Site 81 is for the launch of Proton rockets, and Site 254 is the place to refurbish the Soyuz and Progress spacecraft.

Kevin: Can I visit the astronaut quarters here?

Ivan: Yes, the crew quarters are very famous here. You know there is a tradition that the crewmembers will sign their names on the door of their dormitory when they leave for the launch. You can see those famous signatures.

Kevin: I have heard that cosmonauts have many interesting rituals before liftoff. Is that right?

Ivan: Yes, the most famous one is pee ritual. The cosmonaut will stop on the way to the launch pad and pee at the back tyre of their vehicle.

Kevin: Why?

Ivan: This tradition started in 1961. When cosmonaut Gagarin was on the way to the launch pad, he wanted to go to the toilet badly, so he stopped and peed on the back tyre of his car. This tradition has been kept as a token of good luck ever since.

Kevin: That's so interesting! Thank you!

Kennedy Space Center

Kevin: What is the most famous American launch center?

Ivan: Kennedy space center, of course. It is located at the coast of Florida, about 45-minute drive from Orlando. Kennedy Space Center and Cape Canaveral Air Force Station are working launch facilities which are both there.

Kevin: Is this center open to the public?

Ivan: Yes, you can buy a ticket and tour the center, but some of the areas are restricted.

Kevin: How long does it take to complete a visit?

Ivan: It depends on your itinerary, if you want to have a detailed visit, it might take you up to 8 hours.

Kevin: What sites can I visit in the center?

Ivan: The first site is the main visitor complex. In this place, you can visit the Heroes and Legends building, watch high quality 3D space videos in the IMAX theater, tour the rocket garden, and see the Space Shuttle Atalantis. You can also experience the Journey to Mars and even some astronaut training experiences in this site.

Kevin: Were the Apollo spaceships and the space shuttles launched from this spaceport in the past?

Ivan: Yes, both the Apollo and Shuttle programs were launched from Kennedy Space Center. The Apollo/Saturn V center is open to tourists and you can visit this historic site.

Kevin: What are the main working facilities there?

Ivan: The Vehicle Assembly Building (VAB) is a key facility located in the Launch Complex 39 area. It is one of the largest buildings by volume in the world. It is about 160 meters tall, 218 meters long, and 158 meters wide. It was built for the Apollo program, but was refurbished many times and has been used by the Space Shuttle program, and is still in use now.

Kevin: Where is the firing room in this launch center?

Ivan: It is on the third floor of the Launch Control Center (LCC). LCC is the hub of launch operations at NASA's Kennedy Space Center which locates at the southeast corner of the massive Vehicle Assembly Building.

Kevin: Can you explain more about the firing room?

Ivan: Ok. It was in these firing rooms that the final checkout of each launch vehicle built to carry astronauts was conducted. Once a vehicle received the "go" for launch, the firing room personnel were responsible for the supervision and control of the liftoff and vehicle until it cleared the launch pad towers. After that, the Johnson Space Center in Houston assumed control of the vehicle.

Kevin: What about the launch pads?

Ivan: The most famous launch pads are Launch pads 39 A and B which were a few miles away from LCC. The Apollo spaceships and space shuttles were launched from those pads. And they are still in service. The Demon-2 of SpaceX, the first commercial crew flight also lifted off from this launch pad.

Kevin: Thank you so much for the introduction.

Ivan: You are welcome!

European Spaceport in French Guiana

Kevin: Where is the European spaceport?

Ivan: It is far away from Europe, actually it is in Kourou of French Guiana.

Kevin: Why do they build their launch sites in Kourou?

Ivan: Just like the Hainan launch site, Kourou is located near the equator. You know, the Earth's rotation can act as an extra source of propulsion for the rocket as it brings the satellite into space, saving fuel and money. The selection is also due to its small population, its excellent access to the Atlantic Ocean, and its stable climate.

Kevin: When was the Guiana launch center built?

Ivan: The construction started in the early 1960s. The French government began launching satellites there in 1964. Now the Guiana Space Center is used both by the European Space Agency and the French government to launch satellites into space. And ESA contributes two-thirds of the spaceport's annual budget.

Kevin: How many launch complexes are there in the center?

Ivan: There are mainly four launch complexes, ELA-1, ELA-2, ELA-3, and ELS. ELA-1 is mainly used for Vega rocket, ELA-2 for Ariane-2,3,4, ELA-3 for Ariane-5 heavy rocket, and ELS for Soyuz rocket.

Kevin: Can you introduce some famous launches there?

Ivan: Of course. The Automated Transfer Vehicle (ATV)-the cargo spaceship to resupply the International Space Station, the Rosetta mission, the Herschel and Plank space telescope, the James Webb Space Telescope have all been lifted off from Kourou.

Kevin: Thank you! I really want to visit the center.

Ivan: I believe you can.

Section 3. Astronaut Training Centers abroad

场景介绍：美国约翰逊航天中心、俄罗斯加加林航天员训练中心和欧洲航天员中心是国外著名的航天员训练中心，那里有什么样的训练设施？让我们跟随航天专家一起去"看看"。

Gagarin Cosmonaut Training Center

Ivan: Gagarin Cosmonaut Training Center (GCTC) is one of the most famous astronaut training centers in the world.

Kevin: Where is GCTC in Russia?

Ivan: It is in the Star City, about 25km from downtown Mosco.

Kevin: When was GCTC founded?

Ivan: It was founded in 1960 under the name Air Force Cosmonaut Training Center, and renamed after Cosmonaut Training Center in 1965, and then renamed after Yu. A. Gagarin Cosmonaut Training Center in 1966 to honor Yu. Gagarin. In 1969, the name of the center was changed to Yu. A. Gagarin Research & Test Cosmonauts Training Center, but the abbreviation was still GCTC.

Kevin: What are the responsibilities of GCTC?

Ivan: Of course, the most important job of GCTC is the selection and training of cosmonauts. It is the home base of cosmonaut corps of Russia. Other areas of its work include the medical examination, postflight medical maintenance and rehabilitation, development and modernization of cosmonaut training facilities, as well as scientific researches related to human space flight.

Kevin: So it is not only a training center, but also a medical support and research center! Is the center open to the general public?

Ivan: Yes. You can visit parts of the center as a tourist, such as the Mir Space Station full-scale simulator, TsF-18 centrifuge, the hydro-lab (Neutral Buoyancy Laboratory), and the Museum of the Yu. Gagarin Cosmonaut Training Center.

Kevin: Can you introduce the famous training facilities there?

Ivan: Ok. The unique cosmonaut training facilities are variously dedicated, complex and virtual simulators, such as the ISS Russian Segment Mockup facility and the Soyuz simulators.

Kevin: What are the facilities for weightlessness training?

Ivan: For long term micro-gravity training such as the EVA training, there is the Hydro lab which is 23 m in diameter and 12 m in depth. The mockups of the space station segments can be put in the basin for cosmonaut EVA training. As to short term micro-gravity training or scientific tests, they have an IL-76 MDK flying laboratory. It can create short term micro-gravity conditions during parabolic flights.

Kevin: What else do they have?

Ivan: There are two centrifuges: CF-7 and CF-18. The centrifuges are used for improving the cosmonauts' G-tolerance capabilities.

Kevin: Thank you!

Ivan: You are welcome!

Johnson Space Center

Kevin: Where is the astronaut center of the United States?

Ivan: It is in southeast of downtown Huston and the name of the center is called Johnson Space Center, to honor the late president and Texas native, Lyndon B. Johnson.

Kevin: When was it founded?

Ivan: It was founded and given the name Manned Spacecraft Center in 1961 and renamed as the current name in 1973.

Kevin: Are all the American astronauts living in Huston?

Ivan: Yes, the American astronaut corps is located in Johnson Space Center, so does the international astronauts take training in the center.

Kevin: What training facilities are there in the center?

Ivan: There are many training facilities and most of the American astronaut training are carried out in this Center, for instance, the EVA or spacewalk training in the Sonny Carter Neutral Buoyancy Laboratory, the virtual reality training of EVA or robotic arm operations in Virtual Reality Laboratory, operation training and habitability training in various high-fidelity, full-scale mockups etc.

Kevin: So, there is still some training not taking place in Huston, right?

Ivan: Yes, American astronaut training are quite international, due to the international cooperation nature of the International Space Station. Some trainings are taking place in the international partners, for instance, the Soyuz training in Star City of Mosco Russia, robotic arm training in Canada, Kibo training in Japan, Columbus training in ESA. But that is not all. Some other routine training also take place outside Johnson Space Center in the United States, such as the aircraft training in Ellington Field.

Kevin: How many people are there working in JSC?

Ivan: JSC has a workforce of nearly 10,000 people. You know, astronaut training is only a small part of the responsibility of center.

Kevin: What are the other responsibilities?

Ivan: It is the home to International Space Station mission operations. With a permanent human presence aboard the International Space Station, flight control teams of experienced engineers, medical officers and technicians are on duty 7 days a week, 24 hours a day, 365 days a year.

Kevin: Anything else?

Ivan: It is also dedicated to developing deep space technology and advancing scientific technology. In addition, the White Sands Test Facility located in New Mexico is a satellite of JSC.

Kevin: Thank you for the introduction!

Ivan: You are welcome!

European Astronaut Center

Kevin: Can Europe send astronaut into space?

Ivan: At present, there are only three countries in the world that have the capability of sending humans into space: Russia, the United States and China. But Europe has an astronaut center, and European astronauts have flown into space through international cooperation.

Kevin: Where is the European Astronaut Center?

Ivan: The European Astronaut Centre (EAC) locates in Cologne, Germany. EAC was established in 1990 and it was on the German Aerospace Center (DLR) campus next door to the sate-of-the-art: envihab facility.

Kevin: Is the European Astronaut Corps in EAC?

Ivan: Yes. The European Astronaut Corps is an integration of the astronaut teams of the EAS member states and the ESA core team. The integration started in 1998 and completed in 2002.

Kevin: How many astronauts are there in EAC?

Ivan: Currently, the corps consists of seven active astronauts.

Kevin: Can you introduce the astronaut training in EAC?

Ivan: EAC is not only responsible for the astronaut selection and training, but also for the operations and space medicine. The training of European astronauts is quite international, the EAC astronauts have to take training courses in Russia, in the United States and in other international partners such as Canada and Japan.

Kevin: Why?

Ivan: Because ESA do not have a manned spaceship, so they have to take training courses in Russia if they are going to fly in Soyuz spaceship, or they have to train in the United States if they want to fly in the space shuttle. In addition, the training of robotic arms takes place in Canada, the training of different segments of the ISS in the different countries.

Kevin: What training facilities are there in EAC?

Ivan: There are a range of special training facilities including various replicas of the European Columbus laboratory, stand-alone training models of Columbus' scientific racks, simulators and classrooms, a fully-equipped gym, a 10-meter-deep Neutral Buoyancy Facility, and console rooms for communication with astronauts on board the ISS etc.

Kevin: What are the differences of EAC training as compared with China?

Ivan: Generally speaking they are similar, but the division of training stages of EAC astronauts is different. They mainly categorize the training into basic training, advance training, increment-specific training, EVA training and ATV training.

Kevin: Thank you for the introduction!

Ivan: You are welcome!

Section 4. Terminology and Abbreviations

Astronaut Center of China　中国航天员中心

China Astronaut Research and Training Center

中国航天员科研训练中心

Environment Control and Life Support System

环境控制与生命保障系统

Human factors engineering　人因工程

State Key Laboratory of Space Medicine Fundamentals and Application

航天医学基础与应用国家重点实验室

National Key Laboratory of Human Factors Engineering

人因工程国防重点实验室

Mock-up　舱段模型

Fix-based Spaceship Training Simulator　固定基飞船训练模拟器

Micro-gravity　微重力

Flight procedures　飞行程序

Normal procedure　正常程序

Abnormal procedure　异常程序

Emergency procedure　应急程序

Rendezvous and docking　交会对接

Human centrifuge　载人离心机

Hyper-gravity tolerance　超重耐力

Neutral Buoyancy Laboratory　中性浮力实验室

China Space Station　中国空间站

Extravehicular activity　出舱活动

Residual pressure　余压

Beijing Aerospace Control Center　北京飞行控制中心

"Yuanwang" tracking ship　远望号测量船

Communication coverage rate　通讯覆盖率

"Tianlian" relay satellites　天链中继卫星

Jiuquan satellite launch center　酒泉卫星发射中心

Launch pad　发射台

Vertical assembly, vertical testing and vertical transport

垂直组装、垂直测试、垂直转运

Crew quarter　航天员区

Preflight quarantine　飞行前隔离

Hainan satellite launch center　海南卫星发射中心

Cargo Spaceship　货运飞船

Baikonur Cosmodrome　拜科努尔航天发射场

Sputnik-1　人造卫星 1 号

International Space Station　国际空间站

Soyuz spaceship　联盟飞船

Progress cargo spaceship　进步号货运飞船

Proton rocket　质子火箭

Kennedy Space Center　肯尼迪航天中心

Cape Canaveral Air Force Station　卡纳维拉尔角空军基地

Space Shuttle　航天飞机

Apollo spaceship　阿波罗飞船

Vehicle Assembly Building　航天器组装厂房

Guiana Space Center　圭亚那航天中心

French Guiana　法属圭亚那

Automated Transfer Vehicle　自动转移飞行器

Rosetta mission　罗塞塔任务

James Webb Space Telescope　詹姆斯·韦伯太空望远镜

Gagarin Cosmonaut Training Center　加加林宇航员训练中心

Cosmonaut corps　俄罗斯航天员大队

Mir Space Station　和平号空间站

Johnson Space Center　约翰逊航天中心

Virtual Reality Laboratory　虚拟现实实验室

Kibo　希望实验舱（日本国际空间站舱段）

European Astronaut Centre　欧洲航天员中心

German Aerospace Center　德国宇航中心

European Astronaut Corps　欧洲航天员大队

European Space Agency　欧洲航天局

European Columbus laboratory　欧洲哥伦布实验室

ACC: Astronaut Center of China　中国航天员中心

ATV: Automated Transfer Vehicle　自动转移飞行器

BACC: Beijing Aerospace Control Center　北京飞行控制中心

DLR: German Aerospace Center　德国航空航天中心

EAC: European Astronaut Centre　欧洲航天员中心

EVA: Extravehicular activity　出舱活动

ESA: European Space Agency　欧洲航天局

GCTC: Gagarin Cosmonaut Training Center　加加林宇航员训练中心

NBL: Neutral Buoyancy Laboratory　中性浮力实验室

RVD: Rendezvous and docking　交会对接

VAB: Vehicle Assembly Building　航天器组装厂房

Section 5. Extended Reading Material

1. Speaking to Spacecraft

It may not look like it, but this giant dish in Australia spends its time in in-depth conversation with a number of European deep space missions.

The 35-m antenna is part of ESA's New Norcia ground station, located 140 kilometers north of Perth. The impressive structure is one of three such stations in the Agency's ESTRACK network, designed for communicating with spacecraft exploring the far reaches of the Solar System.

Deep Space Antenna-1 (DSA 1) routinely provides support to Mars Express and Exomars TGO, currently orbiting the Red Planet; the Gaia space observatory, in the process of making the world's most precise map of the stars in our Milky Way galaxy; BepiColombo on its way to Mercury; and Cluster II, studying Earth's magnetic environment.

Preparations are also underway for critical 2020 events, including a crucial BepiColombo flyby and the launch of Exomars RSP and Solar Orbiter.

Discoveries by these spacecraft and others would not be possible without ESA ground stations collecting their data, making it available to researchers across the globe and ensuring we can command and communicate with the explorers from our Operations Centre on Earth.

Vocabulary

antenna：天线

ground station：地面站

ESTRACK network：欧空局跟踪站网络

routinely：例行地

Milky Way galaxy：银河系

BepiColombo：欧空局水星探测器

Mercury：水星

Operations Centre：操作中心

2. Scientists Deepen Understanding of Magnetic Fields Surrounding Earth and Other Planets

Vast rings of electrically charged particles encircle the Earth and other planets. Now, a team of scientists has completed research into waves that travel through this magnetic, electrically charged environment, known as the magnetosphere, deepening understanding of the region and its interaction with our own planet, and opening up new ways to study other planets across the galaxy.

The scientists, led by Eun-Hwa Kim, physicist at the U.S. Department of Energy's (DOE) Princeton Plasma Physics Laboratory (PPPL), examined a type of wave that travels through the magnetosphere. These waves, called electromagnetic ion cyclotron (EMIC) waves, reveal the temperature and the density of the plasma particles within the magnetosphere, among other qualities.

"Waves are a kind of signal from the plasma," said Kim, lead author of a paper that reported the findings in JGR Space Physics. "Therefore, the EMIC waves can be used as diagnostic tools to reveal some of the plasma's characteristics."

Kim and researchers from Andrews University in Michigan and Kyung Hee University in South Korea focused their research on mode conversion, the way in which some EMIC waves form. During this process, other waves that compress along the direction they travel from outer space collide with Earth's magnetosphere and trigger the formation of EMIC waves, which then zoom off at a particular angle and polarization - the direction in which all of the light waves are vibrating.

Using PPPL computers, the scientists performed simulations showing that these mode-converted EMIC waves can propagate through the magnetosphere along magnetic field lines at a normal angle that is less than 90 degrees, in relation to the border of the region with space. Knowing such characteristics enables physicists to identify EMIC waves and gather information about the magnetosphere with limited initial information.

A better understanding of the magnetosphere could provide detailed information about how Earth and other planets interact with their space environment. For instance, the waves could allow scientists to determine the density of elements like helium and oxygen in the magnetosphere, as well as learn more about the flow of charged particles from the sun that produces the aurora borealis.

Vocabulary

electrically charged particles：带电粒子

magnetosphere：磁层

interaction：相互作用

Princeton Plasma Physics Laboratory：普林斯顿等离子体物理实验室

electromagnetic ion cyclotron：电磁离子回旋

plasma：等离子体

lead author：第一作者

diagnostic tools：诊断工具

mode conversion：波形转换

collide with：与…碰撞

polarization：极化

zoom off：放大

propagate：传播

normal angle：法角

aurora borealis：北极光

3. Mars Helicopter Testing Enters Final Phase

NASA's Mars Helicopter flight demonstration project has passed a number of key tests with flying colors. In 2021, the small, autonomous helicopter will be the first vehicle in history to attempt to establish the viability of heavier-than-air vehicles flying on another planet.

"Nobody's built a Mars Helicopter before, so we are continuously entering new territory," said MiMi Aung, project manager for the Mars Helicopter at NASA's Jet Propulsion Laboratory in Pasadena, California. "Our flight model - the actual vehicle that will travel to Mars - has recently passed several important tests."

Back in January 2019 the team operated the flight model in a simulated

Martian environment. Then the helicopter was moved to Lockheed Martin Space in Denver for compatibility testing with the Mars Helicopter Delivery System, which will hold the 4-pound (1.8-kilogram) spacecraft against the belly of the Mars 2020 rover during launch and interplanetary cruise before deploying it onto the surface of Mars after landing.

As a technology demonstrator, the Mars Helicopter carries no science instruments. Its purpose is to confirm that powered flight in the tenuous Martian atmosphere (which is 1% the density of Earth's) is possible and that it can be controlled from Earth over large interplanetary distances. But the helicopter also carries a camera capable of providing high-resolution color images to further demonstrate the vehicle's potential for documenting the Red Planet.

Future Mars missions could enlist second-generation helicopters to add an aerial dimension to their explorations. They could investigate previously unvisited or difficult-to-reach destinations such as cliffs, caves and deep craters, act as scouts for human crews or carry small payloads from one location to another. But before any of that happens, a test vehicle has to prove the possibility.

Vocabulary

autonomous：自主的

viability：可行性

heavier-than-air：比大气重的

project manager：项目经理

flight model：飞行版本

simulated：模拟的

compatibility testing：兼容性测试

interplanetary cruise：行星际巡航

tenuous：稀薄的

scouts：侦察员

4. US military's X-37B space plane spotted in orbit

The US Air Force has been mum about its X-37B programme, drawing the increased attention of the amateur satellite community. The X-37B mini-shuttle is currently conducting its fifth mission, and its planned duration in orbit is still unknown.

An amateur Dutch astronomer has finally photographed the US Air Force's mysterious robotic space plane, the Boeing X-37B, after a months-long hunt.

Ralf Vandebergh, who regularly takes close-up pictures of spacecraft in the Earth's orbit, said he had observed the elusive X-37B on one of its missions.

He said he first saw the space plane in May and then tried to capture it in mid-June, but the X-37B appeared to have escaped his sight after changing its orbit.

"Thanks to the amateur satellite observers-network, it was rapidly found in orbit again and I was able to take some images on June 30 and July 2," the skywatcher wrote, saying that on the most recent occasion, the spacecraft was almost overhead.

The X-37B - otherwise known as the Orbital Test Vehicle (OTV) - looks like a mini-space shuttle, so the image did not show it in much greater detail. "Nevertheless, the images are so successful that they go beyond expectations and show even a sign of the nose/payload bay/tail or even more," Vandebergh said.

"Considering this, the attached images succeeded beyond expectations. We can recognize a bit of the nose, Payload Bay and tail of this mini-shuttle with even a sign of some smaller detail."

But why are some so enthusiastic about spotting a spacecraft?

The point is the US military has been pretty secretive about the mission.

The X-37B OTV is a reusable unmanned space test vehicle. Its first test flight took place in April 2010.

The space plane has been circling the Earth in obscurity for over 20 months on what became the fifth flight of the programme - its longest mission so far.

Vocabulary

mum：沉默不语
mini-shuttle：迷你航天飞机
astronomer：天文学家
elusive：神出鬼没
Orbital Test Vehicle：轨道测试航天器

5. Spaceship Concordia

The 2018 crew of Concordia research station in Antarctica recently returned to the European Astronaut Centre in Cologne to wrap up their time as researchers and subjects at Earth's most remote outpost.

Antarctica has all the wonder and appeal of space. It is harsh, vast and mysterious. But it also has one thing extra going for it: it is a little easier to access.

Peppered throughout the peninsula are research stations like Concordia, a collaboration between the French Polar Institute and the Italian Antarctic programme. Concordia is one of only three bases that is inhabited all year long, and is located at the mountain plateau called Dome C.

As well as offering around nine months of complete isolation, Concordia's location at 3233 m altitude means the crew experience chronic hypobaric hypoxia - lack of oxygen in the brain.

During the Antarctic winter, the crew of up to 15 people also endure four months of complete darkness: the sun disappears from May and is not seen until late August.

Temperatures can drop to -80 C in the winter, with a yearly average of -50 C.

As a station set in Earth's harshest space, Concordia is an ideal stand-in for studying the human psychological and physiological effects of extreme cold, isolation and darkness.

To study how these conditions affect humans, ESA commissions a medical doctor every year to run experiments coordinated by ESA and Concordia partners.

Carmen Possnig, the 2018 medical doctor, followed the effects of lack of sunlight and a less oxygen on herself and her fellow subjects for researchers developing countermeasures to altered motor skills, memory, sleep patterns and moods.

Experiments included playing simple memory games and more complex sessions in simulators requiring subjects to pilot and dock a spacecraft as well as routinely providing blood and urine samples.

Experiments have been proceeding for years, meaning researchers can test effective countermeasures such as light therapy to improve sleep, awareness and mood.

Social cohesion is also vital to the crew at Concordia, as it is on the International Space Station or any other spacecraft that will be heading for space. While addressing physical discomforts from lack of sleep and oxygen deprivation greatly improves mood and morale, the crew must also bond and work as a team in shared isolation. Building a climbing wall and planning festivities helped the 2018 crew.

Safety procedures and training taken months before the crew departs set the tone for crew bonding, much like astronauts training for a mission to the Space Station.

Vocabulary

Concordia research station：协和南极站

appeal：魅力

peninsula：半岛

inhabited：有人居住的

isolation：隔离

chronic hypobaric hypoxia：慢性低压缺氧

harshest：最严苛的

stand-in：替身

psychological and physiological effects：心理和生理效应

simulators：模拟器

light therapy：光疗法

social cohesion：社会凝聚力

oxygen deprivation：缺氧

morale：士气

festivities：庆典

中国航天员科研训练中心 China Astronaut Research and Training Center

中国航天员科研训练中心全景 Panaroma of China Astronaut Research and Training Center

失重训练水槽 Neutral Buoyancy Laboratory (NBL)

飞船模拟器 Spaceship Simulator

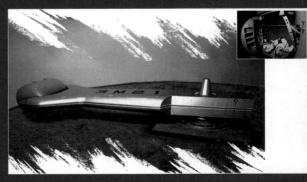

离心机 Centrifuge

北京航天飞行控制中心 Beijing Aerospace Control Center

问天阁 Astronaut Quarter Wentiange

酒泉卫星发射中心发射塔架 Launch Pad
of Jiuquan Satellite Launch Center

酒泉卫星发射中心公园 Park of
Jiuquan Satellite Launch Center

东方红 1 号卫星发射塔架
Launch Pad of Dongfanghong-1 Satellite

海南发射场发射塔架
Launch Pad of Hainan
Satellite Launch Center

海南发射场全景
Panaroma of Hainan Satellite Launch Center

肯尼迪航天中心全景
Panaroma of Kennedy Space Center

肯尼迪航天中心航天器组装厂房
Vehicle Assembly Building of
Kennedy Space Center

拜科努尔发射场
Baikonur Cosmodrome

欧洲法属圭亚那航天发射场
European Space port in
French Guiana

约翰逊航天中心 Johnson Space Center

约翰逊航天中心任务控制中心 Mission Control Center in Johnson Space Center

欧洲航天员中心 European Astronaut Center

哥伦布实验舱训练模拟器 Training Simulator of Columbus Module

加加林航天员训练中心 Gargarin Cosmonaut Training Center

Chapter 2
To Be An Astronaut

航天员分为驾驶员航天员、飞行工程师以及载荷专家／任务专家等类别，不同类别航天员的选拔和训练标准不同。入选的预备航天员要经过近两年的训练才能成为正式航天员。航天员的训练可划分为三个阶段：基础训练、高级训练／专业技能训练、任务训练。本章详细讨论了航天员如何选拔以及航天员的各种训练科目。

Section 1. Astronaut Selection

场景介绍：凯文很想知道怎样才能成为真正的航天员，了解之后才知道"航天员"分为指令长、驾驶员航天员、载荷专家等多种类别，不同类别航天员的选拔标准略有差异。只有通过选拔，才能成为预备航天员。

Astronaut Categories

Kevin: What are the differences among astronaut, cosmonaut and taikonaut?

Ivan: They are synonyms. Astronaut is composed of "astro" and "naut", "astro" means space and "naut" means sailor, or literally space sailor. According to the definition of ASE (Association of Space Explorers), astronaut is the person who has finished at least one orbit space flight. Astronaut is a more general word. Cosmonaut means the Russian astronaut or astronaut aboard the Russian spacecraft. It is also composed of "cosmo" which means cosmos and "naut". Taikonuat is a relatively new word which means Chinese astronaut.It is also composed of "Taiko", which is the phonetics for space, and"naut". These words exactly reflect the nations who are capable of carrying out manned space missions.

Kevin: What types of astronauts do we have?

Ivan: A selected astronaut who has not yet finished training program is called a candidate astronaut. The Americans divide the astronaut into four categories: the former astronaut, the active astronaut, the management astronaut, and foreign astronaut. Former astronaut refers to the retired astronaut who has left NASA. Active astronaut refers to the astronaut who is qualified and can be assigned to a flight mission, while management astronaut means the astronaut who is working in a management position and may be turned into an active astronaut after proper training. Foreign astronaut means the astronaut of another country who is taking training in the U.S. Active astronaut is what we talk about most frequently.

Kevin: So, the astronaut in the astronaut corps must be an active astronaut, right?

Ivan: Yes. But an active astronaut may be a pilot astronaut, a flight engineer, or a mission specialist/payload specialist etc.

Kevin: So complicated! Can you explain in detail?

Ivan: A pilot astronaut is responsible for the operation of the spacecraft. Rich experience in airplane piloting is required for the selection of pilot astronauts. While for a flight engineer, the aviation experience is not mandatory and he or she is more required on science background. A mission specialist or payload specialist is more specialized in a certain mission or a certain payload of a mission.

Kevin: What is a commander of a space mission?

Ivan: Commander is a position of a crew. The commander can make the final decision in the crew and the commander usually comes from the pilot astronaut.

Kevin: What is a space tourist?

Ivan: Space tourist is a person who buys a ticket and travel to space for a certain period of time. There have been several space tourists who have flown to the International Space Station on board the Soyuz spaceship for about 10 days. The expensive ticket for as much as 50 million US dollar is not suitable for the ordinary people. In the future, the cost of space travel will definitely decrease and the ordinary people may soon have the chance to fly into space thanks to the rapid development of commercial space companies and the space tourism.

Kevin: What are the requirements for the space tourists?

Ivan: The requirements for the space tourists will not be as strict as that for the professional astronauts, because they will not be responsible for the operation of the spacecraft. But they do have baseline requirements, and proper training is also required. The space tourists prefer to be called flight participants rather than space tourist, and they argue that they carry out science investigations during their mission and are not just touring in space.

Astronaut Selection

Kevin: I really love space. How can I realize my spaceflight dream?

Ivan: There are two options for you: the first option is to pass the astronaut selection and become a professional astronaut; the second option is to be a successful business person and buy your own ticket to the space.

Kevin: I prefer the first option. What is the process of astronaut selection?

Ivan: When there is an astronaut selection opportunity, it will be announced to the public. Generally speaking, there are several rounds of selections. The first round is usually called the preliminary selection: the documents submitted by the applicants and/or preliminary online tests will be used for a primarily screening and the obviously unqualified applicants will be eliminated. The second round of selection is rather extensive including many tests such as clinical, space medical, and psychological examinations etc. The qualified candidates then will have the chance to enter into the third round of selection—face to face interview. In the end, the proposed candidates will be submitted to the related authorities for final approval. The selection process may last for half a year or even one year.

Kevin: How often do we have the selection opportunity?

Ivan: It depends. Different countries have different arrangements. For the United States, the selection of new class of astronauts is approximately once every two years. But other nations are quite irregular. China selected the first batch of taikonauts in 1998, the second batch in 2010, and the third is expected to be finished in 2020. While for ESA, the latest class of astronauts was selected in 2009.

Kevin: What are the requirements for the astronauts?

Ivan: There are many. The first requirement is nationality, only citizens have the right to take part in the astronaut selection of that country. It is now a general practice for all space-faring nations.

Kevin: What else?

Ivan: Physically you shall be able to endure the stresses encountered during space flight; Psychologically you shall be stable and have good mental compatibility; In addition, you must have proper educational background and at least a bachelor degree in science or engineering, or you should have abundant aviation flight experiences.

Kevin: What does the "stresses during spaceflight" mean?

Ivan: You know when you are taking a space ride, you will encounter various stresses in the process. For instance, during lift-off and landing, there will be acceleration stresses which are normally 3-4g, and can be as high as 8g during ballistic reentry in emergency. There will also be micro-gravity, noise, vibration, confined environment etc.

Kevin: What kind of tests do we have for the endurance of such stress factors?

Ivan: The human centrifuge is used to test the G tolerance. The swing and rotation chair are used to test the vestibular function. The altitude chamber is used to test the hypoxia endurance. The cardiovascular function is also tested because in the environment of micro-gravity, there may be cardiovascular deconditioning. These are just some typical examples of space medical tests.

Kevin: What are the requirements for the height, weight, and eyesight?

Ivan: The requirement on the body height comes from the limitation of space in the manned spacecraft and it is contingent on the specific conditions of each craft. But in the future, with the development of technology, there may be no such restriction as body height. As the body weight is related to the body height, so the more precise parameter is the BMI or body mass index. As to the eyesight, the requirement is based on the corrected visual acuity for the mission specialists.

Kevin: Can you give me some suggestions on how to be an astronaut?

Ivan: You must be good physically and psychologically. You need to exercise regularly to keep fit. You must study hard and get at least a bachelor degree in science or engineering. In addition, you need a bit luck, because the selection of astronaut is so competitive!

Kevin: Thank you for the explanation!

Ivan: You are welcome!

Section 2. Astronaut Training

场景介绍：预备航天员要经过近两年的训练才能成为航天员，航天专家带你"走进"航天员的训练场，深入了解航天员要接受的基础训练、专业技能训练和任务训练中的各种科目。

Training Phases

Kevin: You have told me that the newly recruited candidate is called candidate astronaut, how long will it take to be a qualified astronaut?

Ivan: It will take approximately 2 years.

Kevin: Can a graduated astronaut fly a space mission immediately?

Ivan: No. He or she must make it through specific training for the mission. The training for the mission of astronauts mainly includes three phases: the basic training, the advance training, and the increment training/mission specific training. Mission specific training is the last phase before flight.

Kevin: Can you take the taikonaut as an example and explain the three phases in detail?

Ivan: The training of taikonauts is divided into three phases. First, the candidate astronauts study space technology, basic medical skills, and the overall design of spacecraft and space station. Then they go on into more details about systems of spacecraft and space station, and get the skills of EVA, RVD and robotic arm control. After the first two phases, if assigned to a mission, they will move on to the special tasks linked with the mission. If not, they will learn more about flight procedure, ground control, EVA advanced skills and some other courses to get better preparation for their assignments.

Kevin: How long will it take for a candidate astronaut to be able to fly a mission?

Ivan: It depends on the role in the flight. For the pilot astronaut or flight engineer, it will take at least 48 months training from the very beginning. If he or she is a payload expert, the training period can be much shorter. Anyway, he or she will undergo thousands of hours of training for a space mission.

Training Details

Kevin: Can you tell me more about the courses in the first phase?

Ivan: Of course. The first phase is also called the classic training phase, the astronauts need to take dozens of courses to get basic knowledge on various technical and scientific disciplines. The objective of this phase is to ensure all the candidate astronauts of different professional backgrounds or expertise, to have a common minimum knowledge base in subjects relevant to their career. The astronaut needs to study courses in technical disciplines such as spaceflight engineering, electrical engineering, aerodynamics, propulsion and orbital mechanics, as well as courses in scientific disciplines such as researches in weightlessness(human physiology and biology) , earth observation, astronomy, etc.

Kevin: I see, the astronauts need to learn a lot of knowledge in the basic training phase, do they need to do anything else?

Ivan: Besides the courses, they also need to do a lot of training, such as fitness training, HBP training, space environment adaptation training and so on. The training mentioned above are not related to any specific training phase, so it is also called regular training which the astronaut needs to do every year since recruitment. Some astronauts will do the survival training and scuba diving in the first phase, too.

Kevin: I like fitness training and I spend a lot of time in the gym. Is the astronaut fitness training similar to that?

Ivan: Some of the exercise may be similar, but the objective is quite different. The astronauts do fitness training to develop their physical strength, flexibility and endurance, which helps them endure the acceleration in launch, control their body movement in space, and do a better job in EVA.

Kevin: I see. You have mentioned HBP training, what is that?

Ivan: The full name of HBP training is human behavior and performance training, which aims to help the astronauts get better communication within the crew and improve the collaboration efficiency. ISS astronauts take HBP courses before participating in ESA's CAVES training and NASA's NEEMO training. In China, the training is more about the psychological compatibility training and imagery training which is a little bit different from the HBP training abroad.

Kevin: What is CAVES training and what is NEEMO training?

Ivan: CAVES stands for Cooperative Adventure for Valuing and Exercising human behaviour and performance Skills. CAVES training is organized by ESA and held in Sardinia island in Italy. A crew of astronauts or technicians from different countries are given exploration tasks. The caves they explore is dark, confined and isolated, and they need to practice HBP skills to communicate with each other. NEEMO is also called NASA Extreme Environment Mission Operations, and participants are mainly astronauts, technicians or mission experts. They will form an international crew to live together in the Aquarius Reef Base, located 19 meters below the surface of the Atlantic Ocean off the coast of Florida. The Aquarius habitat and its surroundings provide a convincing analog for space exploration, and the crew need to do tasks such as evaluating equipment for future mission, doing experiments or performing EVAs in the sea. During the process, the crew members need to use HBP skills to keep the communication smooth and effective.

Kevin: I've read news about a taikonaut participating in CAVES training, can you tell me more about it?

Ivan: In 2016, taikonaut Ye Guangfu participated in CAVES training. Ye and 5 international astronauts spent 15 days in the cave together. They explored the unknown area of the cave and found a new branch, and they measured the data and generated a 3D image of the newly-found area. CAVES is about team work, so each crewmember will provide support to others and make decisions together.

Kevin: I believed Ye Guangfu must have achieved a lot in CAVES training.

Ivan: Yes, it is a great experience. After the training, the European Space Agency decided to give the honor to Ye Guangfu to name the newly discovered cave branch due to his excellent performance in the training. Ye Guangfu named it Guangming Gallery, which means gallery of light, because light means hope in the darkness.

Kevin: Thank you for the explanation! Let's come back to training, could you tell me more about the space environmental adaptation training?

Ivan: Of course. Space environmental adaptation training includes centrifuge training, blood redistribution adaptability training, vestibular function training, parabolic flight training, as well as some test experiences of noise, impact, and vibrations in a mission. It helps the astronauts get adapted to the special environment in a manned space flight mission, such as high G load in launch and reentry, weightlessness in orbit, impact of landing and so on.

Kevin: It sounds quite challenging to take part in this kind of training!

Ivan: Yes, it is a great challenge to get through all these training courses. Take the centrifuge training for instance, the centrifuge can create up to 8 times the earth gravity for astronauts. First the instructor will teach the astronauts how to use abdominal muscles to regulate their breath and counteract the effects and tension of high G load. When the instructor has confirmed the astronauts have mastered the skills, the doctors will give astronauts a physical test. If their physical condition is qualified for this training, they will begin training on the centrifuge. On the first run, the load is small and it increases time by time until it reaches 8 G. The astronauts need to keep their mind clear and do operations as instructed while counteracting the high G load.

Kevin: I was told that high G load may cause blackout or even coma. How to make sure the astronauts are safe?

Ivan: There is an emergency button in one hand, if an astronaut feels he or she is going to pass out, he can push the button to stop the centrifuge. The instructors and doctors in the centrifuge control room are also observing the physiological data of the astronaut in the centrifuge cabin, and they would stop the centrifuge if they have found the physiological data is abnormal.

Kevin: Thank you for your explanation, Ivan. I am also confused about blood redistribution adaptability training. Can you tell me more about how to redistribute the blood of astronauts? Is it done by surgery?

Ivan: Okay, Kevin, actually this training has nothing to do with surgery. The astronauts do it on a tilt-bed, while the instructor change the angles of the tilt-bed horizontally upward or downward. In this process, gravity pulls the blood to flow to the feet or head of the astronauts. The angle increases time by time, and the astronaut will get used to the blood redistribution physically.

Kevin: Oh, I see how it works, but why do astronauts take this training?

Ivan: Because the space environmental condition affects the blood distribution of the astronauts. Take the launch of a spaceship for example, the rocket carries the spaceship accelerating upwards after igniting, meanwhile the blood of the astronauts is accumulating downwards because of inertia. Once the astronauts are in orbit, more blood will flow to the head because there is no gravity, and that is why astronauts seem to have rounder cheeks in space. The blood distribution has changed dramatically during the process, and it happens just in less than 20 minutes. So, we need this training to help the astronaut to get used to the blood redistribution.

Kevin: I see, and what is Vestibular function training?

Ivan: The vestibular function training helps the astronauts to get used to the weightless environment in orbit and reduce the probability of space motion sickness. They do it by riding on a rotation chair or swing, which can maintain their vestibular function at a high level.

Kevin: Can you tell me more about the parabolic flight?

Ivan: Okay, normally the astronauts take part in this training only once before assignment. They will be on board a specially modified aircraft which performs a series of parabolas after it reaches certain height. Each flight can do 20~30 parabolas, and each parabola can produce microgravity for 20~28 seconds. For many astronauts, parabolic flight is one of their favorite training experiences, because they can float their body freely in the cabin.

Kevin: It seems appealing to me as well! What do astronauts do while floating?

Ivan: They practice how to control their movement, eat food, drink water, deal with waste, do medical treatment and space experiments without gravity.

Kevin: Where do astronauts take the survival training?

Ivan: We choose different off-road environment for the survival training. The taikonauts have done it in forests, deserts, lakes, or even the sea.

Kevin: Do they need to survive in the wild all by themselves?

Ivan: Yes, they will be divided into groups of two or three to survive in the wild. The astronauts will be given an introduction to the survival skills, equipped with standard survival gear which is the same with that in a spaceship. After that they are sent to the training area, where they begin with egress from a reentry mock-up, and survive in the wild for several days. On the first day, they need to make fire and build shelters with wood and parachute fabric. Before they run out of food and water, they also need to find edible plants and collect clean water. They are also required to send rescue signals to rescuers and navigate to the rescue site with the help of GNSS and compass.

Kevin: Is it possible that astronauts fall into danger in the wild?

Ivan: The training area must be confirmed by experts that it is free of large wild animals and highly poisonous plants or animals in advance. Though we have given lessons about how to keep safe in the wild, it is still possible that the astronauts get hurt or encounter wild animals. Now we use drones to monitor the training area, so normally it is pretty safe. If anyone is hurt, the others in the group need to give emergency treatment and contact the rescuers immediately.

Kevin: You said that the astronaut will also learn scuba diving, is it part of the survival training?

Ivan: No, it is not for survival in the wild. The astronauts need to learn scuba diving before they train in the NBL, or weightlessness simulation water tank. After the training, they will meet the standard of open water diver.

Kevin: Is the EVA training mainly in NBL?

Ivan: Yes. NBL is the primary training site. But the astronaut will first take EVA suit courses to learn how to dress and operate an EVA suit properly.

Kevin: Is it easy to put on an EVA suit?

Ivan: It is not easy at all. First, they put on a diaper, as we know the EVA mission always lasts for hours. Next, they pull on a layer of lightweight underwear and a liquid cooling garment. The liquid cooling garment is a jumpsuit with plastic tubes woven into the garments, which circulate water to keep them cool in EVA suit. Then they will wear the communication headsets so that the others can talk to them when they are in the EVA suit. When they are ready, they come to the EVA suit which is fixed to a rack. The suit weighs over 100 kilograms, so they can't just "put it on", they actually open the backpack and squeeze themselves into it. After they get into the suit, they must check their vision from the helmet, making sure his gloves are attached and locked properly, all the meters displayed are normal, and the communication system works well. Finally, they will close the backpack and check with suit technicians that they are fully clothed.

Kevin: After the EVA suit course, are they going underwater to do EVA training?

Ivan: Not exactly. The astronauts do EVA training in different facilities, and NBL is only one of them.

Kevin: Really? I only know about the NBL, can you tell me more about other facilities?

Ivan: Of course. We have EVA procedure training simulator, weightlessness simulation water tank(NBL) and a VR simulator.

Astronauts use EVA procedure training simulator to get familiarized with EVA procedure of egress and ingress through the airlock. The training in NBL is to master how to move in weightless environment along the handrails on the surface of the cabin or on the end of a robotic arm. They also practice all the operations of EVA mission underwater. The VR simulator can give the astronaut a simulated view of the deep space and whole space station in orbit. They put on VR goggles and use controllers to do all their operations in the three-dimensioned weightless world displayed.

Kevin: How long does it take for the astronaut to complete the EVA training?

Ivan: It depends on the EVA mission they are about to carry out. Basically, they need to spend hundreds of hours in training to develop their EVA skills.

Kevin: I can tell from EVA training that simulator is very important to astronaut training. Is that true?

Ivan: Absolutely. There are many simulators and mock-ups in ACC, such as space station, space ship and cargo ship simulators and mock-ups, RVD simulator, robotic arm simulator and so on. They play a key role in astronaut training.

Kevin: Do the astronauts learn how to pilot a spacecraft in spaceship simulator?

Ivan: Yes. After they have studied the technical manuals of Shenzhou spaceship, they will begin training the spaceship simulator, which can simulate the sound, view and parameters of Shenzhou spaceship from launching to landing. The astronauts can use the RVD control handles in the simulator to rendezvous and dock onto the space station. They can also practice dealing with malfunctions in spaceship systems and emergency reentry. The training instructor will serve as the "ground" to communicate with them, and give technical support when necessary. When they have finished training on spaceship simulator, they are familiarized with the flight procedure and master all the skills needed to fly on a spaceship!

Kevin: It sounds cool to control a spaceship, but it is so difficult to be an astronaut, maybe I can experience the flight in the simulator someday.

Ivan: That's a good idea! But if you want to experience the astronaut's life in orbit, maybe you should come to the space station simulator.

Kevin: Why?

Ivan: Because the space station is astronauts' home in orbit. The simulator is a full-scale replica of the space station, providing as much realism as possible to match the conditions the astronauts will experience onboard. In this simulator, the astronauts need to practice taking care of themselves in space, such as preparing food and drinks, disposing waste, maintaining personal hygiene, sleeping, utilizing physical exercise gears, monitoring personal health condition and so on. They also need to practice operating and maintaining the systems of the space station, such as operations of all the panels, logistic management of the supplies from cargo ship, public sanitation, system malfunction treatment, and so on. They also receive training in emergency evacuation and laboratory systems familiarization in the simulator. So, if you want to experience their daily life in space, it is the best choice.

Kevin: If an astronaut has done all the training above, is he or she ready for a flight mission?

Ivan: Almost, but he has to do one last training course at the launch site before launching, and we call it preflight rehearsal. The day before the rocket propellent loading, the flight crew are dressed in pressure suits and enter the spaceship capsule. The preflight rehearsal is 1:1 similar to the real launch, and all the systems involved in launching will participate in it. It starts from about 3 hours before the launch, and all the systems need to coordinate with each other, making sure everything is ready as required. The crew need to check whether the device and equipment is in good condition inside the capsule, the air-tightness of pressure suit is qualified and the communication with all the systems smooth and clear. If everything goes well with the preflight rehearsal, the crew will be fully ready for their flight.

Kevin: Thanks for your explanation. And I've got an overall understanding of astronaut training.

Ivan: You are welcome, that's all for the training session.

Section 3. Terminology and Abbreviations

Astronaut　航天员

Cosmonaut　俄罗斯航天员

Taikonaut　中国航天员

Association of Space Explorers (ASE)　太空探索者协会

candidate astronaut　预备航天员

active astronaut　现役航天员

management astronaut　管理航天员

pilot astronaut　驾驶员航天员

flight engineer　飞行工程师

mission specialist/payload specialist　任务专家 / 载荷专家

commander　指令长

Space tourist　太空游客

flight participant　飞行参与者

ballistic reentry　弹道再入

confined environment　狭小环境

hypoxia endurance　缺氧耐力

cardiovascular deconditioning　心血管失适应

body mass index(BMI)　身体质量指数

corrected visual acuity　矫正视力

basic training　基础训练

spaceflight engineering　航天工程

aerodynamics　空气动力学

fitness training　体质训练

space environment adaptation training　航天环境适应性训练

survival training　生存训练

scuba diving　水肺潜水

Human Behavior and Performance training (HBP training)

人类行为与绩效训练

Cooperative Adventure for Valuing and Exercising human behaviour and performance Skills(CAVES)

洞穴训练（评估与实践人类行为与绩效技能的联合探险）

NASA Extreme Environment Mission Operations（NEEMO）

美国航空航天管理局极端环境任务操作

Aquarius habitat　水瓶座居住舱

blood redistribution adaptability training　血液重新分布训练

vestibular function training　前庭功能训练

parabolic flight training　抛物线飞行训练

blackout　黑视

coma　昏迷

survival training　生存训练

liquid cooling garment　液冷服

EVA procedure training simulator　出舱程序训练模拟器

VR simulator　虚拟现实模拟器

egress and ingress　出入

spaceship simulator　飞船模拟器

preflight rehearsal　飞行前演练

ASE: Association of Space Explorers　太空探索者协会

BMI: body mass index　身体质量指数

CAVES: Cooperative Adventure for Valuing and Exercising human behavior and performance Skills

洞穴训练（评估与实践人类行为与绩效技能的联合探险）

HBP: Human Behavior and Performance　人类行为与绩效

NEEMO: NASA Extreme Environment Mission Operations

美国航空航天管理局极端环境任务操作

Section 4. Extended Reading Material

1.Explorers Wanted: NASA to Hire More Artemis Generation Astronauts

As NASA prepares to launch American astronauts this year on American rockets from American soil to the International Space Station – with an eye toward the Moon and Mars – the agency is announcing it will accept applications from March 2 to 31 for the next class of Artemis Generation astronauts.

Since the 1960s, NASA has selected 350 people to train as astronaut candidates for its increasingly challenging missions to explore space. With 48 astronauts in the active astronaut corps, more will be needed to crew spacecraft bound for multiple destinations and propel exploration forward as part of Artemis missions and beyond.

"We're celebrating our 20th year of continuous presence aboard the International Space Station in low-earth orbit this year, and we're on the verge of sending the first woman and next man to the Moon by 2024," said Administrator Jim Bridenstine of NASA. "For the handful of highly talented women and men we will hire to join our diverse astronaut corps, it's an incredible time in human spaceflight to be an astronaut. We're asking all eligible Americans if they have what it to takes to apply from beginning of March 2."

The basic requirements to apply include United States citizenship and a master's degree in a STEM field, including engineering, biological science, physical science, computer science, or mathematics, from an accredited institution. The requirement for the master's degree can also be met by:

- Two years (36 semester hours or 54 quarter hours) of work toward a Ph.D. program in a related science, technology, engineering or math field;

- A completed doctor of medicine or doctor of osteopathic medicine degree;

- Completion (or current enrollment that will result in completion by June 2021) of a nationally recognized test pilot school program.

Candidates also must have at least two years of related, progressively responsible professional experience, or at least 1,000 hours of pilot-in-

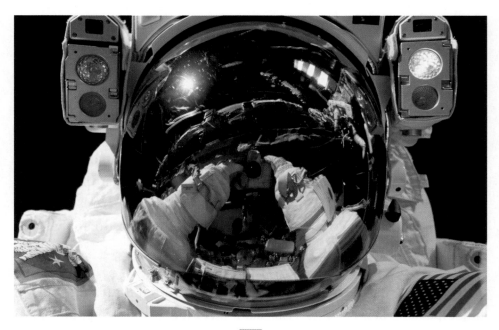

command time in jet aircraft. Astronaut candidates must pass the NASA long-duration spaceflight physical.

Vocabulary

application：申请

Artemis：阿尔特弥斯（希腊神话中的月亮与狩猎的女神）

astronaut candidates：预备航天员

active astronaut：现役航天员

bound for：驶往

low-Earth orbit：地球低轨道

astronaut corps：航天员大队

eligible：符合条件的

STEM：科学、技术、工程、数学

accredited institution：得到认证的机构

osteopathic medicine：骨科医学

test pilot：试飞员

2. From 'Cavewalking' to Spacewalking

It might not be obvious, but there are many similarities between working deep underground and in outer space.

Just as with spacewalks, underground 'cavewalks' require safety tethering, 3D orientation, careful planning and teamwork. Cave explorers need to stay alert in an environment where they are deprived of natural light and every move is a step into the unknown.

The ESA's CAVES training course has been taking astronauts below the earth's surface and prepared them to work safely in an environment where the terrain, climate and climbing techniques pose high demands.

Jeanette Epps, a NASA astronaut, is seen here, to the left, hanging over 200 m of void and closely monitored by certified speleology instructor, Marco Vattano, during her descent. After a week of preparations, Jeanette explored a cave in Slovenia where she lived and worked for six days with five other 'cavenauts'.

Jeannette Epps follows another another NASA astronaut Jessica Meir, who became the first woman to participate in CAVES in 2016 and recently starred in an all-female spacewalk outside the International Space Station.

The procedure for moving along a cave wall strongly resembles spacewalking. "You have to be very aware of the position of your body and what is around you. If you hit something or miss a step, the consequences are critical," explains a NASA astronaut Mike Barratt in this video about his CAVES adventure back in 2013.

As a veteran spacewalker, Mike Barratt points out how working in darkness along a handrail and using safety tethers all times is quite similar to walking in space in the Russian Orlan spacesuit. Managing stress and taking decisions in an alien environment for long period is exhausting.

As with a launch and landing in a spacecraft, entering and exiting from a cave are the most critical moments. Caves are dangerous environments and the explorers could face situations where the intervention of speleologists with advanced technical skills could be required to prevent accidents.

Vocabulary

outer space：外太空

safety tethering：安全系绳

orientation：定向

deprived of：剥夺

CAVES training course：洞穴训练课程

terrain：地形

void：空间

speleology instructor：洞穴学教员

cavenauts：洞穴员

resemble：类似

adventure：探险

veteran：经验丰富的

handrail：扶手

Orlan spacesuit：奥兰航天服

alien environment：陌生环境

3. Thomas Pesquet on a new underwater lunar adventure

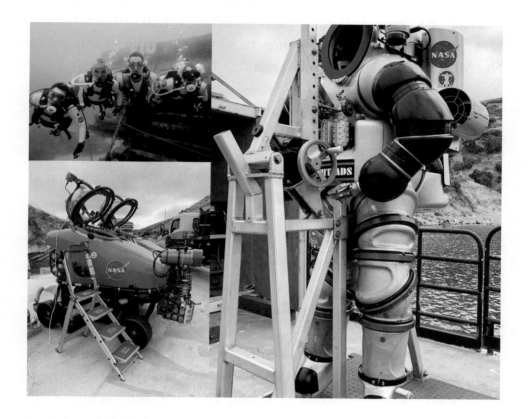

Over the next two weeks ESA astronaut Thomas Pesquet, NASA astronaut Drew Feustel and Japanese astronaut Norishige Kanai will take part in a new NASA Extreme Environment Mission Operations (NEEMO) mission off the coast of California, USA, to assess concepts for undersea training that will aid astoronauts next steps on the Moon.

Testing new technology, tools, techniques and training programs for space exploration starts on Earth, and space agencies head to extreme environments to put their ideas into practice, simulating aspects of space.

ESA organizes regular missions in caves in Italy and on the Canary Islands, while NASA has run 23 NEEMO missions off the coast of Florida, USA. Thomas Pesquet participated in NEEMO 18 in 2014 and ESA astronaut Samantha Cristoforetti was commander of this year's NEEMO sortie that took place last month.

This expedition is investigating a new concept for NEEMO with greater focus on exploration and a return to the Moon in 2024. Unlike previous NEEMO missions, the astronauts, also known as aquanauts, will not live underwater but resurface each day to stay on the island of Santa Catalina, an island off the coast of southern California.

Thomas Pesquet and Drew Feustel will test submersibles and underwater suits to judge whether the new location is suitable as a stand-in for the environment astronauts will encounter on the Moon. Thomas Pesquet is assigned as submersible pilot for the mission.

The underwater world offers similar geological features to the Moon and putting ideas to the test under high workload, real-world stress, and the unexpected problems that come with field work is one of the only ways to iron out any kinks in new exploration approaches and techniques.

Thomas Pesquet was chosen to take part in this feasibility study due to his background as an airline pilot and spacewalk experience on the International Space Station during his Proxima mission in 2016. He also took part in ESA's CAVES program alongside Norishige Kanai, as well as two other NASA underwater missions, making him an ideal astronaut to assess the new program.

Support divers and ground-based personnel will monitor the tasks and guide the astronauts. Experts from ESA's astronaut center and Japan Aerospace Exploration Agency (JAXA) are taking part too.

Vocabulary

Extreme Environment Mission Operations (NEEMO)：极端环境任务操作

undersea training：海底训练

simulating：模拟

sortie：出击任务

aquanauts：海底实验工作人员

submersibles：潜水器

underwater suits：水下服

geological features：地质学特征

iron out：消除

approache：方法

airline pilot：航空公司飞行员

support diver：提供支持的潜水员

ground-based：基于地面的

4.Space Adventures, Inc. Announces An Agreement With SpaceX To Launch Private Citizens On The Crew Dragon Spacecraft

Building on the success of Crew Dragon's first demonstration mission to the International Space Station in March 2019 and the recent successful test of the spacecraft's launch escape system, Space Adventures, Inc. has entered into an agreement with SpaceX to fly private citizens on the first Crew Dragon free-flyer mission. This will provide up to four individuals with the opportunity to break the world altitude record for private citizen spaceflight and see planet Earth the way no one has done since the Gemini program.

If interested parties are secured, this mission will be the first orbital space tourism experience provided entirely with American technology. Private citizens will fly aboard SpaceX's fully autonomous Crew Dragon spacecraft launched by the company's Falcon 9 rocket, the same spacecraft and launch vehicle that SpaceX will use to transport NASA astronauts to the International Space Station (ISS).

"This historic mission will forge a path to making spaceflight possible for all people who dream of it, and we are pleased to work with the Space Adventures' team on the mission," said Gwynne Shotwell, President and Chief Operating Officer, SpaceX.

"Creating unique and previously impossible opportunities for private citizens to experience space is why Space Adventures exists. From 2001-2009 our clients made history by flying over 36 million miles in space on eight separate missions to the ISS. Since its maiden mission in 2010, no engineering achievement has consistently impressed the industry more than the Dragon/Falcon 9 reusable system. Honoring our combined histories, this Dragon mission will be a special experience and a once in a lifetime opportunity - capable of reaching twice the altitude of any prior civilian astronaut mission or space station visitor," said Eric Anderson, Chairman, Space Adventures.

Vocabulary

Crew Dragon：载人龙飞船

launch escape system：发射逃生系统

Space Adventures：太空探险公司

private citizens：普通公民

free-flyer：自由飞人

space tourism：太空旅游

Falcon 9 rocket：猎鹰 9 火箭

Chief Operating Officer：首席运营官

maiden mission：首次任务

reusable system：可重复使用系统

5. Fit for space – survival training

In the third episode of our 'Fit for space' training series, the ESA's astronaut Matthias Maurer explains how astronauts learn to survive on their own if ever a spacecraft lands away from its intended landing site.

Generally, a spacecraft lands within a few kilometers of its landing site, but sometimes they return in a so-called 'ballistic mode' in a steeper entrance trajectory putting the astronauts under increased gravity loads, and landing with less precision – sometimes 400 km away from the intended landing area.

From winter survival to water survival, astronauts are prepared for anything, from building fires and shelters to surviving cold waters and righting capsized life rafts.

The video includes footage from the Soyuz winter survival training, water survival training fire emergency training as well as Chinese sea survival training.

Vocabulary

episode：一集

survive：生存

landing site：着陆地点

ballistic mode：弹道模式

entrance trajectory：再入轨道

gravity loads：重力负载

winter survival：冬季生存

water survival：水上生存

footage：影片镜头

emergency training：应急训练

Chinese sea survival training：中国海上生存训练

中国 11 位飞天航天员合影 Group Photo of 11 Chinese Taikonauts

美国 2017 届航天员 NASA 2017 Class Astronauts

模拟器训练 Simulator Training

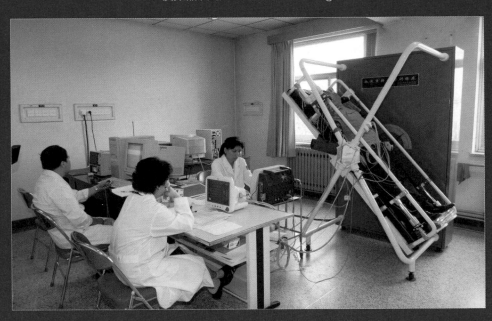

血液重新分布分布训练 Blood Redistribution Training

转椅训练
Rotation Chair Training

中国航天员参加俄罗斯失重飞机训练
Taikonauts Trained onboard Russian Zero-g Plane

失重飞机 Zero-g Plane

失重飞机训练 Parabolic Flight Training

航天员出舱训练 EVA Training

海上救生训练 Sea Survival Training

直升机吊救 Helicopter Rescue

沙漠生存训练 Desert Survival Training

体质训练 Fitness Training

航天员出洞后与所有工作人员合影
Group Photo of All CAVES
Astronauts and Experts

艰难的通过狭窄通道
Crawling across the Cave

海洋模拟环境训练 NEEMO Training

Chapter 3
Flying into Space

本章聚焦进入太空的过程。首先讨论了飞行乘组任务前的医学隔离，航天飞行前各种有趣的仪式，发射倒计时以及进入太空的过程；然后系统介绍了运载火箭和各种载人航天器，包括各国的载人飞船、已退役的航天飞机以及空间站等，并介绍了为航天员进行补给的各种货运飞船。

Section 1. Before Launch

　　场景介绍：载人航天器发射之前，航天员要进行周密细致的准备。发射前两周，航天员就要开始进行医学隔离。航天专家将为我们讲解飞天前有趣的仪式、发射倒计时的场景，以及载人航天器进入太空的整个过程。

Quarantine of Astronauts

Kevin: What preparations will an astronaut make before flying into space?

Ivan: They will prepare themselves in almost every aspects of their mission. They will take training continuously and maintain good health. A special arrangement before flight is quarantine.

Kevin: That's interesting! Because of the COVID-19 pandemic, we are quite familiar with the terms like quarantine and social distancing. Why do astronauts have the quarantine?

Ivan: That's a good question! Living in space is quite different from living on the earth. The microgravity environment will weaken the immune system of astronauts. While pathogens in space can reproduce faster, and make things worse, they might become more deadly in space. So, we have to protect the astronauts from being sick in space.

Kevin: When will the astronaut quarantine begin?

Ivan: It starts two weeks before the flight mission.

Kevin: Why?

Ivan: Because the incubation period of most infectious diseases is 10-14 days. If the astronauts are infected, the symptoms will show up during the quarantine period. If there is no symptom, we are sure the astronauts are free from infections. So, the astronauts will not take potential health risks while flying in space.

Kevin: What shall we do if an astronaut is sick during the quarantine?

Ivan: The flight doctors will evaluate the health status of the astronaut. If he or she is not able to carry out the space flight mission. He or she will be replaced by the backup. You know, for every manned space mission, there are always the primary crew and the backup crew.

Kevin: I know that the astronauts will take training during the quarantine, then how to effectively implement the quarantine measures?

Ivan: Actually, the close contacts of the crew will take the quarantine together with the astronauts, such as the instructors and the medical support staff. If occasional contacts are needed, for instance, the media interview, the persons involved will follow the related quarantine procedures, such as the medical examination and certification of that person, keeping a certain distance with the astronaut, etc.

Kevin: What's your comment on the quarantine measure?

Ivan: It is a very important and rather effective. In the early stages of the human space flight when the quarantine measures were not adopted, severe health incidents happened again and again in space. Some missions had to be terminated and the astronauts had to return early due to the illness! But after the adoption of the quarantine, it never happened again!

Kevin: Thank you for the introduction!

Ivan: You are welcome!

Interesting Rituals before Flight

Kevin: I have heard that astronauts have many interesting rituals before they fly into space, is that right?

Ivan: Yes. Many traditions have been formed and inherited in human space flight which may be traced back to Soviet cosmonaut Yuri Alekseyevich Gagarin in 1961.

Kevin: Can you give me some examples?

Ivan: Of course. Planting trees by the crew several days before the flight is a long tradition. In the Avenue of Heroes of Baikonur, a city in Kazakhstan, you can see the trees planted by various cosmonauts including the tree planted by Gagarin nearly 60 years ago. In China, taikonauts also plant trees before the flight mission in Jiuquan Satellite Launch Center.

Kevin: Anything else?

Ivan: Yes. In the morning of the launch day, the crew are given champagne at breakfast and then they sign their names on the door of their dormitory. If you have the chance to visit the crew quarters in the launch sites such as Jiuquan Satellite Launch Center or Baikonur Cosmodrome, you can see the signatures of astronauts which can give you a very good sense of history.

Kevin: You have told me previously that cosmonauts will urinate on their way to the launch pad.

Ivan: Yes, that is a very unusual ritual for Russian cosmonauts. The ritual can be dated back to April 12, 1961 when Gagarin was on his way to the launch pad. All of a sudden, he wanted badly to go to toilet. He stopped and peed on the back right wheel of his vehicle. After the success of the first human space flight, peeing on the wheel has been kept as a tradition and it means good luck. Even today, the cosmonauts or astronauts flying onboard Soyuz from Baikonur still keep this tradition. Males will pee on the back right wheel, while females will collect urine in a small bottle and spray on the wheel. You know this tradition is only limited to cosmonauts flying Russian spaceship, other space-faring nations like China and the United States do not do that.

Kevin: Can you tell me something about the blessings?

Ivan: Ok. Blessing is a relatively new tradition in Russia. An Orthodox Russian priest blesses the rocket once it has reached the launch pad. The priest also blesses the crew the day before the launch, sprinkling them with holy water.

Kevin: Some rituals seem a little superstitious!

Ivan: Yes, but this is not surprising, considering human space flight involves sitting atop massive rocket fuel and being blasted away from your home planet.

Kevin: Any other superstitious rituals?

Ivan: In Russia, the crew is not supposed to watch the roll-out of the rocket and capsule to avoid bad luck. While engineers, support staff and their families will place coins along the track when the capsule is being pulled along to the launch pad by a train. They believe it could invite good luck to the mission.

Kevin: What about the cuddly toy of astronauts?

Ivan: The cuddly toy on a chain is dangled to the instrument panel and in full view of the cameras. Actually, the toy is served as a zero-g indicator. When the spacecraft enters into micro-gravity, the toy will float and we know immediately that the crew enters into orbit.

Kevin: Thank you for the introduction!

Ivan: You are welcome!

Countdown

Kevin: When do the crew enter the capsule on the launch day?

Ivan: Actually, the crew enters the capsule hours before the lift-off. The crew will be very busy during the launch day, medical examinations, preparations, donning the pressure suit, and departure ceremonies. Once in the capsule, there will be fewer interferences and the crew can finally concentrate on their works.

Kevin: Can astronauts take their personal things with them?

Ivan: Yes, the astronauts are permitted to take certain weight of personal stuff into space, such as family photos, books, souvenirs, snacks, etc.

Kevin: Why do the astronauts wear the pressure suit during launch?

Ivan: The pressure suit or IVA(Intravehicular Activity) space suit is used to ensure the safety of astronauts. In case of depressurization of the space capsule, for instance, the spaceship is hit by a micrometeorite and the air is leaking, the pressure suit can maintain the normal air pressure and supply oxygen for the astronaut. In Soyuz-11 mission, three cosmonauts were killed during reentry in 1971 due to depressurization accident. After that, wearing the pressure suit became mandatory for astronauts during the launch, RVD, and landing operations.

Kevin: Why do the astronauts lie down on their seats during launch?

Ivan: That's a very good question! During lift-off and landing, the astronauts will experience strong accelerations. Taking supine position can help the astronauts to better cope with the G-loads.

Kevin: How long does it take to enter into the orbit?

Ivan: It takes about 10 minutes.

Kevin: How to rescue the crew during launch if a fatal accident happens?

Ivan: For the spaceship, if the rocket is still on the launch pad, the astronauts can take the escape tunnel and retreat to the underground shelter. If the rocket has left the launch pad, the engines of the escape tower on top of the rocket can take the space capsule to a safe place and thus save the lives of astronauts.

Kevin: Once in space, do they still have to wear their pressure suits?

Ivan: No, they can wear normal working suits.

Kevin: Thank you for the introduction!

Ivan: You are welcome!

Section 2. Launch Vehicles and Spacecraft

场景介绍：载人航天器由火箭发射入轨，凯文向专家兴致勃勃地请教了许多知识。其中包括各种航天运载工具、不同国家的载人飞船、已经退役的航天飞机，以及人类的太空家园——空间站。

Rocket

Kevin: What kind of rockets do we use for human space flight in China?

Ivan: You know, we have liquid rockets and solid rockets. For human space flight missions, liquid rockets are usually used, though some launch vehicles may also use solid boosters, for instance, the space shuttle. But the key point is the safety and reliability, the reliability of human-rated rocket is much higher than that of the unmanned rocket.

Kevin: What kinds of fuel are used in the rocket?

Ivan: The rocket carries both the propellent and the oxygen. The new generation rockets use environment friendly propellent such as liquid hydrogen or kerosene. But the traditional rocket propellent is highly toxic.

Kevin: Why is the rocket divided into stages?

Ivan: The reason is to transport more payloads into orbit with the same amount of propellent. For instance, when the fuel of the first stage is used up, the first stage will be separated and jettisoned. So, the fuel of the rest stages will be used only for the rest of the rocket and it will be more efficient.

Kevin: Where is the spaceship located in the rocket?

Ivan: It is in enclosed in the fairing on top of the rocket.

Kevin: Why?

Ivan: When the rocket is in the atmosphere, the high speed will destroy the spaceship if there is no fairing to provide aerodynamic protection.

Kevin: What is the little tower on top of the manned rocket?

Ivan: It is called the escape tower which is used for emergency. It can bring the spaceship to a safe place even if the rocket explodes after lifting off from the pad. However, in the latest crew dragon spaceship of SpaceX, the escape tower is replaced by thrusters in the space capsule.

Kevin: Why?

Ivan: You know the commercial company is always trying to decrease the cost. Generally speaking, the escape tower will be jettisoned when everything is normal. The commercial company believes that is a big waste.

Kevin: It seems reusable rockets is quite attractive?

Ivan: Definitely! The SpaceX company has already realized that. Some recovered rockets have been used for five times. That significantly decreased the cost of space launches. You know the fuel is not expensive at all. Many counties are studying the related technology of reusable rocket.

Kevin: Why do we have the blast deflector in the launch pad?

Ivan: Without the deflector, the rocket may be destroyed by its own flames.

Kevin: Thank you for the introduction!

Ivan: You are welcome!

Spaceship

Kevin: If I want to fly to space, what kind of spacecraft can I take?

Ivan: The most traditional method of space travel is by spaceship.

Kevin: Can you tell me how the spaceship works?

Ivan: Yes. The spaceship is launched vertically into space by rocket. It is equipped with life support system which can create a habitable environment for astronauts. When the spaceship returns to the Earth, it will make a parachute-assisted vertically landing.

Kevin: Is the life support system in the spaceship the same with that on the space station?

Ivan: No, the life support system in the spaceship is non-regenerative, so spaceship can only support short duration flights. But the life support system on the space station is regenerative, that is to say, the oxygen and water are mostly recycled which significantly reduces the resupply demands of the space station.

Kevin: I have read that only the return module of a spaceship comes back to the Earth. Is that right?

Ivan: Yes. Generally speaking, a spaceship consists of a service module, a return module and an orbit module. The service module provides power and propellent. The return module is in the middle of a spaceship where the astronauts operate the spaceship, and it is the only part that will return to the Earth. The orbit module is for the living of astronauts, it is connected to the return module and has a docking mechanism at the other end. But in some new spaceships, the return module and the orbit module are integrated into one module.

Kevin: How to get the power supply in a spaceship?

Ivan: The spaceship has solar panels which can convert the sunlight into electrical power.

Kevin: What spaceships are used to transport the crew to the International Space Station?

Ivan: Currently, the Russian Soyuz-MS is the only manned spaceship to service the ISS, but the Crew Dragon of SpaceX is testing its crew transportation under Demo-2 mission, and in the near future the commercial company SpaceX is expected to be able to provide crew transportation services for the ISS.

Kevin: Can you introduce the Crew Dragon?

Ivan: The Crew Dragon spacecraft is developed by SpaceX under NASA's Commercial Crew Program. Crew Dragon is launched atop the Falcon 9 rocket and can take a maximum of 7 astronauts. The falcon 9 rocket and the spacecraft are reusable which significantly reduces the cost.

Kevin: Can you introduce other spaceships of the United States?

Ivan: Well, in the past, NASA has developed the Mercury spaceship, the Gemini Spaceship and the Apollo Spaceship. The purpose of the Mercury is to make a breakthrough of crew transportation technology, while the Gemini is to test the rendezvous and docking technology. The Mercury and the Gemini are the basis for the development of the Apollo spaceship which eventually realized the manned lunar landing in 1969. Now, all those spaceships are retired.

Kevin: What are the current manned spaceships in the United States?

Ivan: In the US, the Crew Dragon is the closest to be put into service. Moreover, NASA is also developing its own spaceship Orion which is targeted for deep space exploration including the Moon and the Mars. Boeing is developing the Starliner spaceship, also under the Commercial Crew Program of NASA, but the progress is not satisfactory and the latest test flight is only partially successful.

Kevin: What about the Russian spaceships?

Ivan: In the past, the Soviet Union/Russia has developed the Vostok, Voskhod, and Soyuz spaceships. Soyuz-MS is the latest version of Soyuz spaceship and the only active spaceship of Russia.

Kevin: Can you introduce the Chinese spaceships?

Ivan: China is the third country that can send humans into space. China has developed Shenzhou spaceship which can carry three taikonauts. It adopts a three-module configuration, the service module, the return module and the orbit module. You know China has recently tested its new generation spaceship which is a two-module configuration and can accommodate 4-7 taikonauts.

Kevin: Can you explain more of the new Chinese spaceship?

Ivan: It is not only more spacious and takes more taikonauts, it will also be reusable. The newly adopted group parachutes and airbag cushioning will make the spaceship landing more comfortable.

Kevin: Is there any other country developing manned space capabilities?

Ivan: India is reported to plan to send its own astronauts into space.

Kevin: Thank you for the introduction!

Ivan: You are welcome!

Space Shuttle

Kevin: What is the difference between the space shuttle and the spaceship?

Ivan: The space shuttle is launched vertically just like the spaceship, but the landing of space shuttle is more like a commercial airline, this is significantly different from the vertical landing of a spaceship. In addition, the space shuttle is designed to be reusable. The orbiter of the space shuttle is designed to be reused for over a hundred times after maintenance. But most of the spaceships in the past were not reusable.

Kevin: How many astronauts can a space shuttle carry?

Ivan: The transportation capability of space shuttle is much bigger than the spaceship, it can carry up to 7 astronauts and over 30 tons of payloads. While the spaceship in the past could only carry 3 astronauts and only limited amount of cargos.

Kevin: Can you introduce the configuration of the space shuttle?

Ivan: The space shuttle includes an orbiter, two solid boosters and an external tank. Only the orbiter enters into space and returns to Earth after mission. The solid boosters are recovered and reused, so is the orbiter. The payloads are mounted in the payload bay of the space shuttle.

Kevin: The space shuttle is reusable and can send more crew and cargo into space, why did they retire the shuttle fleet?

Ivan: That's a very good question! The space shuttle has a illustrious history, but it had its own problems and the shuttle fleet was completely grounded in 2011. The most important reason was safety. The space shuttle is extremely complex and has used many "new" technologies which decreased its reliability and safety. Since the maiden shuttle flight in 1981, two shuttles have been lost, Challenger exploded in 1986 during lift-off and Columbia disintegrated in 2003 during reentry. Another reason is the cost. Although space shuttle is reusable, the maintenance fee is about 0.5 billion dollars once which is 10 times higher than expected. It is even more expensive than the cost of a brand-new single use spaceship.

Kevin: How many space shuttles have been built?

Ivan: The first prototype of space shuttle is Enterprise, but it is only for test, so not included in the fleet. The United States has built five space shuttles: Columbia, Challenger, Discovery, Atlantis, and Endeavour.

Kevin: I have heard that the Soviet Union has also built a space shuttle, is that right?

Ivan: Yes, the Soviet Union has developed a space shuttle, and the name of the shuttle is called Buran, but it has only had an unmanned flight test. Then the program was cancelled.

Kevin: What are the main achievements of the space shuttle?

Ivan: The space shuttle has played a very important role in manned space flight of the United States. The fleet has flown many space lab missions, deployed many famous payloads like the Hubble Space Telescope, performed on-orbit satellite retrieval and servicing. And the most important thing is that it has played a vital role in the construction of the International Space Station, many large segments like the Destiny Module, the Columbus Module, the Japanese Experiment Module Kibo were all transported by the space shuttle.

Kevin: Where was the shuttle launched?

Ivan: The space shuttle used to be launched from launch pad 39A in Kennedy Space Center.

Kevin: Can they fly like an airliner?

Ivan: The space shuttle has a limited maneuver capability, if the weather is not favorable during landing in Kennedy Space Center in Florida, then it can land in the Edwards Air Force Base in California. But from Edwards Air Force Base, the space shuttle has to be transported on top of a dedicated aircraft to Kennedy Space Center.

Kevin: Where are the retired shuttles now?

Ivan: They are in the museums of the Unites States. If you visit Kennedy Space Center you can see the space shuttle Atlantis.

Kevin: Thank you for the introduction!

Ivan: You are welcome!

Space Station

Kevin: What is the difference between the spaceship and the space station?

Ivan: Spaceship is like a car taking astronauts to and from space, while space station is like a house where astronauts can live.

Kevin: So, the space station must be much bigger than spaceship, right?

Ivan: Yes, the space station usually consists of multiple modules and it is thus much more spacious than the spaceship. For instance, the pressurized volume of the International Space Station (ISS) is over 900 cubic meters and the weight is over 400 tons.

Kevin: Is ISS the only space station in the Low Earth Orbit (LEO)?

Ivan: Yes, ISS is the only human habitat in space now, but it is expected to be retired in 2024. China is going to finish the construction of its own space station in 2022. Maybe after 2024, the only space station in LEO will be the China's Space Station.

Kevin: How old is ISS?

Ivan: The first module of ISS was launched in 1998 and has been continuously occupied by astronauts since November 2000.

Kevin: What space stations have been built in the past?

Ivan: After the moon race between the United States and the Soviet Union, the U.S. turned to the space shuttle program and the Soviet Union turned to the space station program. In the past, the Soviet Union has launched 7 Salyut space stations and a Mir Space Station. The Mir space station is the third-generation space station and had been operated in space for over 15 years from 1986 to 2001.

Kevin: What are the main components of the ISS?

Ivan: Comparing with the Mir space station, ISS is the fourth-generation space station. The solar panels attached to the truss provide power for the station. The space manipulator or the robotic arm on the ISS supports the construction of ISS, the crew transportation during EVA, and the grabbing of the cargo spaceship, etc. There are modules such as FGB, Service Module, Destiny, Columbia, and JEM experiment modules which are the main parts of ISS. The three nodes serve as junctions to connect different modules. The airlocks on ISS are used for EVA of astronauts. While the docking ports are for berthing of the visiting space vehicles. You see, the ISS is very complex.

Kevin: How many countries have participated in the ISS program?

Ivan: The ISS is an international collaboration of many countries. The main players include NASA(The United States), ROSCOSMOS(Russia), ESA(The European Union), JAXA(Japan), and CSA(Canada).

Kevin: How many astronauts can ISS accommodate?

Ivan: The ISS can support a crew of 6 astronauts to work there, although during crew rotation there may be more astronauts for a short period of time.

Kevin: What facilities are there on ISS?

Ivan: The research facilities are numerous. Outside the space station, there are exposure platforms. Inside the station, there are large amount of science racks for various disciplines in the experiment modules. The researches on ISS mainly include biological sciences and biotechnology, human research, physical sciences, earth and space science, as well as technology demonstrations.

Kevin: Can you introduce the life support system on ISS?

Ivan: The life support system on ISS is physical-chemical regenerative type. Most of the water and oxygen are recycled. You know, the water and oxygen supply of 6 astronauts will be massive if not recycled. Most of the food are supplied from the ground. In the future, if we want to establish a human habitat on the Moon or Mars, the food shall also be produced in situ. That will rely on the controlled ecological life support system (CELSS) which is extensively studied on ISS.

Kevin: The orbit of ISS will decrease due to the drag of trace air in space, how to maintain the orbit?

Ivan: The cargo spaceships docked to the ISS will be used to elevate the orbit of ISS.

Kevin: Thank you for the introduction!

Ivan: You are welcome!

Cargo Transportation

Kevin: How to send cargo to the space station?

Ivan: Cargo spaceship is the main tool to resupply goods to the space station. Although, the manned spaceship has limited capability of transporting goods to and from space.

Kevin: What is the main difference between a cargo spaceship and a manned spaceship?

Ivan: The cargo spaceship is pressurized, but the cargo spaceship does not have the life support system and other human facilities like the escape tower and the thermal protections.

Kevin: What cargo spaceships are used to resupply the International Space Station?

Ivan: The most frequently used cargo spaceship is the Russian Progress. It can send about 2.5 tons of cargo including water, oxygen, propellent and scientific instruments to the ISS. JAXA's H-II Transfer Vehicle

(HTV), ESA's Automatic Transfer Vehicle (ATV), the commercial Dragon and Cygnus can also transport cargo to the ISS.

Kevin: Where are those cargo spaceships launched?

Ivan: The Russian Progress is launched from Baikonur by Russian Soyuz rocket, HTV from Tanegashima Space Center in southern Japan atop the H-II rocket, ESA's ATV from European space port Kourou by Ariane 5 rocket, the commercial Dragon cargo spaceship of SpaceX from Cape Canaveral Air Force Station in Florida by Falcon 9 rocket, and another commercial Cygnus cargo spaceship of Orbital ATK from NASA's Wallops Flight Facility in Virginia by Antares rocket.

Kevin: What are the differences of these cargo spaceships?

Ivan: The first aspect of difference is the transportation capacity, the capacity of Progress and Cygnus is about 2.5 tons, the capacity of HTV and Dragon is about 6 tons, while ATV can send up to 7 tons of cargo. The second aspect is the docking method, some cargo spaceship like Progress can dock with ISS automatically, some cargo spaceships like Cygnus, HTV, ATV do not have the docking mechanism and will be captured and berthed by astronauts with the robotic arm.

Kevin: How long will a cargo spaceship dock to the ISS?

Ivan: It will dock to the ISS for 1-2 months. During this period of time, the astronauts will unload the cargo from the ship and re-fill the ship with trashes produced on ISS. You know most of the cargo spaceships will burn up in the atmosphere during reentry. The only cargo spaceship that can return to Earth is Dragon. In addition, the cargo spaceships can be used to elevate the orbit of the ISS when needed.

Kevin: Does China have cargo spaceships?

Ivan: Yes. China has successfully tested its Tianzhou cargo spaceship. This is very important for the construction of China Space Station. Space transportation technology, EVA technology, and RVD technology are the three basic technologies of space station. China has mastered all these three technologies.

Kevin: Can you explain more about the reentry of the cargo spaceships?

Ivan: The cargo spaceship usually does not have the thermal protection. It will perform a controlled reentry at the end of the mission. Due to the high velocity, the cargo spaceship loaded with trash will burn up in atmosphere over sparsely populated ocean areas.

Kevin: Can you introduce the biggest cargo spaceship ATV?

Ivan: ESA has produced a total of 5 ATVs and now ESA has stopped its production. But the ATV technology is still playing a role. The service module of Orion manned spaceship of NASA is made by ESA based on ATV technology.

Kevin: Thank you for the introduction!

Ivan: You are welcome!

Section 3. Terminology and Abbreviations

incubation period　潜伏期

primary crew　主乘组

backup crew　备份乘组

close contacts　密切接触者

space-faring nation　航天国家

capsule　太空舱

Intravehicular Activity　舱内活动

Micrometeorite　微陨石

supine position　仰卧位

solid booster　固体火箭助推器

human-rated rocket　载人火箭

fairing　整流罩

escape tower　逃逸塔

non-regenerative　非再生式

regenerative　再生式

solar panel　太阳能面板

Commercial Crew Program　商业乘员计划

Mercury spaceship　水星座飞船

Gemini Spaceship　双子星座飞船

Orion spaceship　猎户座飞船

Starliner　星际航班

Vostok　东方号飞船

Voskhod　上升号飞船

Orbiter　轨道器（航天飞机）

Payload　有效载荷

Challenger　挑战者号

Columbia　哥伦比亚号

Discovery　发现号

Atlantis　阿特兰蒂斯号

Endeavour　奋进号

Buran　暴风雪号

Hubble Space Telescope　哈勃太空望远镜

Destiny Module　命运号实验舱

Low Earth Orbit　地球低轨道（近地轨道）

space manipulator　太空机械臂

physical-chemical regenerative　物理 - 化学式再生

in-situ　原位（就地）

Controlled Ecological Life Support System　受控生态生命保障系统

H-II Transfer Vehicle H-II　转移飞行器

Cygnus　天鹅座

Tanegashima Space Center　种子岛航天中心

Cape Canaveral Air Force Station　卡纳维拉尔角空军基地

Orbital ATK　美国轨道科学公司

Wallops Flight Facility　沃洛普斯飞行设施

CELSS: Controlled Ecological Life Support System　受控生态生命保障系统

CSA: Canada Space Agency　加拿大航天局

HTV: H-II Transfer Vehicle H-II　转移飞行器

IVA: Intravehicular Activity　舱内活动

ISS: International Space Station　国际空间站

JAXA: Japan Aerospace Exploration Agency　日本航天局

JEM: Japanese Experiment Module　日本实验舱

LEO: Low Earth Orbit　地球低轨道近地轨道

NASA: National Astronautics and Space Administration

美国国家航空航天管理局

ROSCOSMOS: Russian Federal Space Agency　俄罗斯联邦航天局

Section 4. Extended Reading Material

1. Skylab

Liftoff of Skylab 1 came on May 14, 1973, but within minutes it was apparent that there was trouble. William Schneider, NASA's Skylab Program Manager, filled in the details at a post-launch news conference.

"At approximately 63 seconds into the launch of Skylab 1, there was an indication of premature deployment of the meteoroid protective shield," Schneider said. "If that has happened, the shield was probably torn off. The thermal indications are that it is gone, and we have some indication that our solar array on the workshop also did not fully deploy."

As a result of the uncertainty, launch of Skylab 2 with the crew of

Charles Conrad, Joseph Kerwin and Paul Weitz, scheduled for the next day, was postponed.

The NASA-industry team around the United Stafes of America went into action to develop plans and hardware necessary to save Skylab. The astronauts practiced using special tools to remove material that jammed the remaining solar array to allow it to provide Skylab with the needed electrical power. A square thermal shield, which operated like a sunshade, also was developed to protect the station from the heat of the sun.

The crew launched on May 25, 1973 aboard an Apollo command-service module and mission commander Charles Conrad expressed confidence that their preparations would pay off right away.

"This is Skylab 2, we fix anything," he said at the moment of liftoff.

The crew deployed the new solar shield through a small scientific experiment airlock, located in the side of the workshop normally facing the sun. Once outside, the shield popped open like a parasol, with four struts extending outward from a segmented center post. Temperatures inside the lab soon diminished to near-normal levels.

Next came the spacewalk to free the jammed solar away. After considerable work, Joseph Kerwin was able to cut the metal that had jammed the solar wing in a folded position. Using a rope sling, Conrad forced the array beam to deploy. Full extension of the solar panel occurred finally, providing electrical power crucial for the three planned piloted missions.

With the Skylab 2 mission back on track, the crew focused on the primary goals of the program -- studies in materials processing in microgravity, Earth observations, expanding knowledge of solar astronomy, and proving that humans could live and work in space for extended periods.

Vocabulary

liftoff: 发射

premature deployment: 过早展开

meteoroid protective shield: 微流星防护板

torn off: 撕破

jammed: 堵塞

thermal shield: 热屏蔽

parasol: 阳伞

rope sling: 吊索

2. Cargo load

The Cygnus NG-12 cargo vehicle hangs out after arriving to the International Space Station on November 4th 2019.

The latest resupply mission includes over 4 tons of science experiments, crew supplies, and station hardware. It also crucially includes components essential for the series of spacewalks taking place in November, 2019.

In a few weeks ESA astronaut Luca Parmitano and NASA astronaut Drew Morgan will venture out to perform a series of spacewalks four years in the making. The extravehicular activities, or EVAs, will service and enhance the dark matter-hunting Alpha Magnetic Spectrometer

AMS-02.

The dark-matter hunter was launched in 2011 and records over 17 billion cosmic rays, particles and nuclei a year. Results from the particle physics detector are among the top five most-cited publications from International Space Station research.

The instrument was initially edsigned to run for only three years but has been so successful that its mission has been extended. However, three of the four cooling pumps have stopped functioning and required multiple spacewalks to repair.

Luca Parmitano would take a leading role in the spacewalks with the first intended to determine just how and where to intervene, and what tools were needed for the process.

Vocabulary

Cygnus：天鹅座飞船

cargo vehicle：货运飞船

crew supplies：乘组补给

components：部件

nuclei：原子核

most-cited publications：引用最多的出版物

cooling pumps：冷却泵

3. Manned Spacecraft Soyuz MS-13 Completes Redocking Between ISS Modules

A manned Soyuz MS-13 spacecraft and its crew have completed redocking from the International Space Station (ISS)'s Zvezda module to the Poisk module, according to a broadcast conducted by Roscosmo, Russian space agencys on its website.

The manoeuvre was aimed at freeing up the docking port on the Zvezda module for a repeated attempt to dock a Soyuz MS-14 spacecraft with humanoid robot Skybot F-850, FEDOR inside, scheduled for 03:12 GMT on Tuesday. The Soyuz MS-14 failed to dock with the ISS on Saturday due to a malfunction of a signal amplifier on the Kurs automated rendezvous system for the Poisk module's docking port.

The redocking was completed at 03:59 GMT.

The crew of the Soyuz MS-13 during the operation included Russian cosmonaut Alexander Skvortsov, Italian astronaut Luca Parmitano of the European Space Agency, and NASA astronaut Andrew Morgan.

The operation to redock the spacecraft was carried out by Skvortsov in manual mode, and the crew was wearing Sokol spacesuits for safety reasons.

The Soyuz MS-14 with humanoid robot Skybot F-850, nicknamed FEDOR, was successfully launched into space on Thursday, August 22nd, 2019. the spacecraft, however, failed to dock at the ISS on schedule and started retreating from the station. The spacecraft was later pulled to a safe distance from the ISS.

Apart from Skvortsov, Parmitano and Morgan, the ISS crew currently comprises Russian cosmonaut Alexey Ovchinin and NASA astronauts Christina Koch and Nick Hague.

Vocabulary

Redocking：再次对接

Zvezda module：星辰舱

Poisk module：探索舱

Manoeuvre：机动

Humanoid robot：类人机器人

automated rendezvous：自动交会

manual mode：手动模式

nicknamed：绰号

retreating：撤离

4. SpaceX Cargo Spacecraft Splashes Down in Pacific Ocean with Scientific Research

SpaceX Dragon cargo spacecraft carrying 4,200 pounds of scientific experiments and other cargo back to Earth departed the International Space Station at 12:01 p.m. EDT Monday, and splashed down in the Pacific Ocean at 5:48 p.m. (2:48 p.m. PDT), June 3rd, 2019.

Flight controllers at mission control in Houston used the space station's Canadarm2 robotic arm to detach Dragon from the Earth-facing side of the station's Harmony module and maneuver the vehicle into its release position. Expedition 59 Flight Engineer David Saint-Jacques of the Canadian Space Agency monitored its departure as the spacecraft was released through ground-controlled commands.

Dragon's thrusters fired to move the spacecraft a safe distance from the station before SpaceX flight controllers in Hawthorne, California, commanded its deorbit burn. The SpaceX recovery team retrieved Dragon packed with science samples from human and animal research, biology and biotechnology studies, physical science investigations and education activities.

NASA and the Center for the Advancement of Science in Space (CASIS), the nonprofit organization that manages research aboard the U.S. national laboratory portion of the space station, received time-sensitive samples and worked with researchers to process and distribute them within 48 hours of splashdown.

Dragon, the only space station resupply spacecraft currently able to return to Earth, launched on May 4th, 2019, on a SpaceX Falcon 9 rocket from Space Launch Complex 40 at Cape Canaveral Air Force Station in Florida, and arrived at the station two days later for the company's 17th NASA-contracted commercial resupply mission to the station.

Vocabulary

cargo spacecraft：货运飞船

depart：离开

splashed down：溅落

mission control：任务控制

robotic arm：机械臂

detach：脱离

Harmony module：和谐号舱

thruster：推进器

deorbit：离轨

retrieve：回收

time-sensitive：时间敏感的

resupply spacecraft：补给飞船

5. No-fly boys: new Russian space suit clashes with pee ritual

Russia on Thursday unveiled a new space suit for a future spaceship, but the design may have to be changed to continue a decades-old tradition: making a stop to pee on the way to the launch.

The Sokol-M prototype suit was demonstrated at an airshow outside Moscow, as a future replacement for suits currently worn during launches to the International Space Station on Soyuz spacecraft.

The Soyuz, in use since the 1960s, is set to be phased out and replaced in the next few years with a new Russian ship called the Federation.

Zvezda, an aerospace firm and the maker of the suits, says they will be made of "new materials" and adaptable to different body sizes. The current suits must be custom-made for each individual.

But the new design makes it impossible to carry on one particular ritual launched by the first man in space Yury Gagarin, who had to stop and relieve himself on the back wheel of the bus that was taking him to the launch pad in 1961.

The stop has been replicated at almost every launch from the Baikonur launch pad and many male cosmonauts and astronauts pee on the tire for good luck -- something that would be impossible in the new suit, according to its maker.

Women astronauts are not obligated to participate but some have been known to bring vials of their urine to splash on the tire instead.

The bright orange Sokol-M suit sports one diagonal zipper, rather than the currently used white Sokol-K, which has a V-shaped opening pointing to the crotch area.

The new suit is lighter due to the zipper and all seams being airtight, so there is no longer a need for the internal rubberized layer known as "the bladder" which is difficult to get in and out of.

All members of the ISS crew currently have to wear the made-to-measure Sokol-K suits, since they must be worn on the Soyuz spacecraft that brings cosmonauts and astronauts to orbit.

The suits are made to be comfortable while reclining in the deep Soyuz chairs, but are heavy and contribute to a noticeable stoop when walking up to the rocket.

The Soyuz launches are full of rituals, in which cosmonauts plant trees at Baikonur, get haircuts, watch the Soviet classic "White Sun of the Desert" and have an Orthodox priest fling holy water on them.

Vocabulary

Unveiled：揭幕

decades-old tradition：几十年的历史传统

Sokol-M prototype：鹰 -M 样机

Airshow：航展

Soyuz spacecraft：联盟飞船

phased out：逐步淘汰

Federation：联邦

Adaptable：可适应

custom-made：定制

ritual：仪式

replicated：重复

vial：小瓶

diagonal zipper：斜拉链

crotch：胯部

seam：缝合

airtight：气密的

made-to-measure：定做

reclining：倾斜

stoop：弯腰

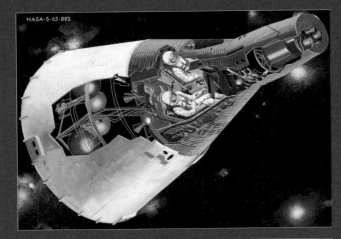

双子星座飞船
Gemini Spaceship

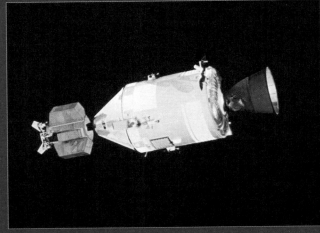

阿波罗飞船
Apollo Spacecraft

联盟飞船飞行中
Soyuz in Space

神舟 11 号飞船发射 Launch of
Shenzhou-11 Spaceship

飞船反推发动机点火 Thruster Firing
during Spaceship Landing

联盟飞船发射 Launch of Soyuz

联盟飞船逃逸塔 Escape Tower of Soyuz

航天飞机发射 Launch of Space Shuttle

航天飞机在 NASA 改装的 747 飞机上
Space Shuttle Atop Its Modified NASA 747 Carrier Aircraf

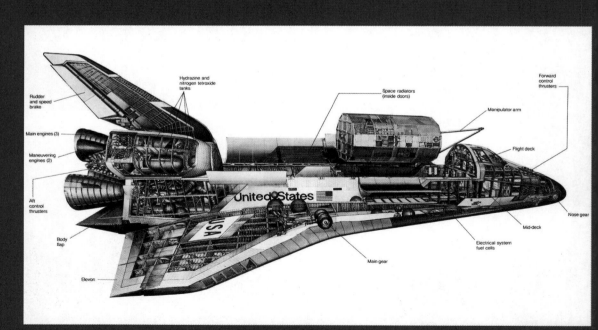

航天飞机轨道器结构图 Illustration of Space Shuttle Orbiter

俄罗斯进步号货运飞船
Progress Cargo Spaceship

欧空局 ATV ESA ATV

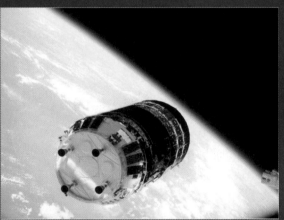

日本 HTV Japanese HTV

礼炮—1 试验性空间站
Salyut-1 Experimental Space Station

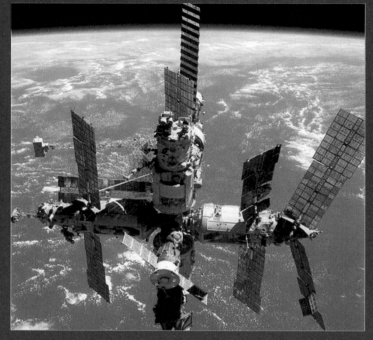

和平号空间站 Mir Space Station

航天飞机对接和平号空间站
Space Shuttle Docked with Mir Space Station

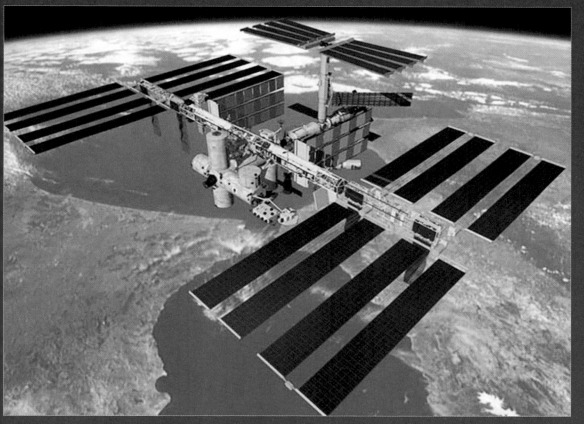

国际空间站 International Space Station

隔离 Quarantine

飞行前祈福 Blessings before Flight

俄罗斯航天员飞行前签名
Door Signing by Cosmoanut before Flight

航天员签名
Astronaut Signatures

航天员植树 Tree Planting by Astronauts

中国首次载人航天飞行出征仪式
Ceremony of the First Manned Spaceflight of China

Chapter 4
Living in Space

　　太空是一个真空、微重力环境，阳照面与阴照面温差达四百摄氏度，此外还有辐射、微陨石等不利因素。利用环境控制与生命保障技术，创造适合人居的人工环境，人类才能在太空生存。

　　载人航天主要包括地球低轨道探索和深空探索。太空生活和工作同地面相比，有着巨大差别。长期驻留太空的航天员会发生肌肉萎缩、骨质流失等一系列生理变化，他们要通过体育锻炼等对抗措施，在医学监督与医学保障团队支持下，维持身心健康。航天员在太空中的工作既有舱内的各种科学实验、载人平台的维护、个人的生活照料，也包括充满危险的出舱活动。

　　航天员的返回主要包括陆地着陆和海上溅落，长期飞行后，航天员需要一个半月左右的时间进行恢复。本章对话，将对上述内容进行了详细解答。

Section 1. Space Environment and Human Survival

场景介绍：太空是与地面截然不同的高真空、微重力、高辐射环境。航天专家带你"走进"太空，了解如何在太空保障人类的生存。

Space Environment

Kevin: Can you briefly explain the space environment?

Ivan: Space is a harsh environment where human cannot survive if there is no proper protection.

Kevin: How harsh is the space environment?

Ivan: It is a vacuum and there is no air in space. If a human being is exposed to vacuum without any protection, the body fluid will boil and he or she will explode and die immediately.

Kevin: That is terrible! I have learned that space is a weightless environment, is that right?

Ivan: Yes. Weightlessness is more frequently called "micro-gravity". The term "micro-gravity" is more accurate than weightlessness, because there is trace atmosphere in space which will produce tiny resistance force and there will be tiny "gravity". Micro-gravity is an induced environment where the spacecraft orbiting the Earth experiences a constant free-fall. Micro-gravity is also sometimes called "zero gravity".

Kevin: Gravity does not disappear in space; it just offsets the centrifugal force of the orbiting spacecraft! Is that right?

Ivan: Yes. Gravity between two objects is proportional to their masses and inversely proportional to the square of their distance. The gravity does not disappear, but is used to pull the spacecraft from flying away.

Kevin: Can we create micro-gravity on the ground?

Ivan: Yes, we can create short term micro-gravity on the ground. Airplanes can be modified to zero-G planes. By flying parabolic flights, the zero-G planes can create 20-30 seconds micro-gravity in each parabola, each sortie can fly dozens of parabolas. We can also use drop tower to create short term micro-gravity. But we cannot create long term micro-gravity on the ground.

Kevin: Let's come back to the space environment, what else is different from the ground?

Ivan: The thermal environment is quite different in space. When facing the sun, the temperature will be as high as over one hundred degrees Celsius, while when in shielding the temperature can be lower than minus two hundred degrees Celsius. Heat transfer in space is mainly by radiation, there is no convection and few conductions.

Kevin: What does that mean to us?

Ivan: The harsh thermal environment imposes a big challenge to the materials used in spacecraft. They will have to endure such a harsh thermal environment.

Kevin: Is there radiation in space?

Ivan: Yes, the radiation in space is much higher than that on the ground. First, I'd like to make an explanation about the spectrum of the electromagnetic wave, you know all the visible, invisible lights or radiations are electromagnetic waves. For the wavelength from longer to shorter, we have the radio, microwave, infrared, visible, ultraviolet, X-ray, and gamma ray. When we talk about radiation, we mainly refer to the harmful short wavelength spectrum radiation, or ionizing radiation.

Kevin: What is ionizing radiation?

Ivan: It refers to the types of radiation that can cause ionizations, as it passes through a material. Ionizing radiation includes X-rays and gamma rays in the electromagnetic spectrum, as well as various heavy particles.

Kevin: How to measure the radiation?

Ivan: The radiation dose can be expressed in units of Sievert (Sv). Radiation is harmful to the health of human body, there are limitations for the maximum exposure dose and accumulated exposure dose in life time. Within the limitation, the human is safe.

Manned Space Exploration

Kevin: The space environment is not suitable for the survival of human being.

Ivan: Yes. If we want to travel to space, we must create a habitable environment inside a spacecraft. The key of the habitable environment is the Environmental Control and Life Support System(ECLSS). That is the reason why manned spacecraft is much more complicated than the unmanned spacecraft.

Kevin: Can you explain the ECLSS?

Ivan: ECLSS can create and maintain a habitable environment for human in space. It can maintain the normal air pressure, humidity, and temperature in the cabin, provide oxygen supply, carbon dioxide removal, and waste disposal.

Kevin: Can we recycle the wastes in ECLSS?

Ivan: That is a very good question. ECLSS has evolved from the non-regenerative to the regenerative. For short duration flights, we can use the non-regenerative type of ECLSS; but for long duration flights, we must use the regenerative type of ECLSS to recycle the wastes. On the space station, physical-chemical regenerative ECLSS is usually used. In the future, Controlled Ecological Life Support System will be used especially for the Lunar or Mars habitats in manned deep space explorations.

Kevin: If we travel to space, what are the main destinations?

Ivan: Space flights can be roughly divided into LEO and deep space exploration missions.

Kevin: What is a LEO?

Ivan: LEO is the abbreviation of Low Earth Orbit. It is an Earth-centered orbit with the altitude less than 2,000 km and the eccentricity less than 0.25.

Kevin: What kinds of spacecrafts are in LEO?

Ivan: Most scientific satellites and many weather satellites are in a nearly circular, Low Earth Orbit. The recent manned spacecrafts are all in LEO including the International Space Station, and the to-be-built China Space Station.

Kevin: What is deep space exploration?

Ivan: The term is used for the exploration of deep space and which is usually described as being quite distant away from the Earth, within or away from our solar system. Actually, Lunar or Mars missions belong to deep space exploration.

Kevin: What is the difference between the LEO and deep space exploration?

Ivan: In LEO, the radiation is much smaller than the deep space exploration, because the magnetic field of the Earth can shield most of the radiations. In addition, there is almost no communication delay in LEO, but for deep space exploration the communication delay can be a big barrier for the timely ground support.

Kevin: What is the magnetic field of the Earth?

Ivan: You know some planets have a magnetic field that acts like there is a giant bar magnet in the center of a planet. The magnetic field can be aligned differently with the rotational axis. For example, the Earth's magnetic field is tilted by about 18° with respect to our rotation axis.

Kevin: How does the magnetic field protect radiation?

Ivan: The magnetic field can shield the energetic, charged particles coming from the Sun and other places of the universe. When charged particles run into a magnetic field, they will be deflected and spiral around the magnetic field lines.

Kevin: What is communication delay?

Ivan: Because the distance is so big, even the communication signal travelling at 300,000 kilometers per second takes a long time to reach. The communication delay is quite significant in deep space explorations. For instance, the communication delay for Earth-Moon is 1.3s, Earth-Mars is 3~21 minutes, Earth-Jupiter is 33~53 minutes, Earth-Pluto is 5 hours and Earth-Nearest Star is 4 years!

Kevin: What about the ground support of the deep space exploration?

Ivan: The ground support of deep space exploration is rather limited. Because of the distance, the crew cannot be evacuated to the ground timely when there is an emergency. Due to the communication delay, the crew will have to make decisions on their own. The crew will be most likely autonomous.

Kevin: How to provide logistics to the deep space exploration?

Ivan: It is very difficult. The logistics can be either sent in advance or carried by the crew themselves. So, the in-situ resource utilization technology is very important, that will help solve the logistics problem.

Kevin: Thanks for the explanation!

Ivan: You are welcome!

Section 2. Living in Space

场景介绍：太空生活并不是看上去的那么惬意，航天专家带我们了解太空环境给航天员身体带来的挑战，在太空如何维护航天员的健康和工作效率，以及航天员在太空环境中要完成的各种工作。

Real Space Life

Kevin: Living in space seems so comfortable and cosy!

Ivan: Yes, the astronauts in space looks cosy, but actually it is not! Living in micro-gravity has significant adverse effects on the health of astronauts. They will experience the fullness of the head, space motion sickness, bone loss, muscle atrophy, etc.

Kevin: Can you explain them in detail?

Ivan: Without the force of gravity to pull body fluids downward, the fluids of the body tend to shift upward when living in space. That will cause the fullness of head and stuffy nose. That is why the astronauts look blush and swelling in the face. Fluid shift can cause a series of physiological changes.

Kevin: What is space motion sickness?

Ivan: It is quite similar to the motion sickness on the ground, the symptoms include headache, nausea, sweating, dizziness, etc. It usually happens in the first few days in space and nearly 60-80% space travelers will experience some sort of symptoms.

Kevin: Why do they have the space motion sickness?

Ivan: There are several hypotheses and the most widely accepted hypothesis is the sensory conflict theory, that is, the conflict between what we see and what we feel induces the space motion sickness.

Kevin: How to deal with it?

Ivan: If the symptom is severe, we can use medicine such as Phenergan to cure it. In addition, the astronaut is advised not to move the head suddenly in space, which can decrease the stimulation on the vestibular organ. That can effectively reduce the incidence of space motion sickness. You can observe that the astronaut will turn his whole body instead of the head at the beginning of the flight.

Kevin: What is bone loss and why?

Ivan: In micro-gravity environment, the human body do not need to deal with the gravity as that on the ground. Due to the unloading of the gravity, the minerals of the bones will decrease continuously and the bones become brittle and more likely to fracture, this is quite similar to the osteoporosis on the ground.

Kevin: What is the muscle atrophy and why?

Ivan: On the ground, we must constantly use certain muscles to support ourselves against the force of gravity. In micro-gravity environment, the disuse of these anti-gravity muscles will result in a significant decrease of their mass. The loss of muscle mass corresponds to a loss of strength that can be potentially dangerous if an astronaut must perform a strenuous emergency procedure upon re-entry into the Earth's gravitational field.

Kevin: I am told that the astronauts will be taller in space, is that right?

Ivan: Yes, the astronauts will be taller for several centimeters in space, because the absent of gravity can make the vertebral column elongate and straighten. Unfortunately, the height of astronaut will come back to normal when he or she returns to the Earth.

Kevin: What are the consequences of physiological changes in space?

Ivan: The human body is a very delicate system. Those changes are the results of body adaptation to the micro-gravity environment. If the astronauts live in space for ever, those changes are good. But if the astronauts return to the Earth later, those changes will be a great barrier for them to re-adapt to the 1G environment. So, countermeasures must be taken to mitigate those adverse effects.

Countermeasures to Micro-gravity

Kevin: What countermeasures do we have to maintain the health of astronauts?

Ivan: The answer to this question comes from the science of space medicine which studies the underlying mechanisms of body changes induced by space environment and the countermeasures.

Kevin: Can you explain the countermeasures?

Ivan: Exercise can effectively improve the bone loss and muscle atrophy. Exercise equipment is very important for long duration manned space flights. There are various types of exercise devices such as the vibration free treadmill, the bicycle ergometer, the elastic band for resistance exercise, etc. Astronauts in long duration flights are required to exercise for at least 2.5 hours a day, six times a week while in orbit.

Kevin: How about the effects of exercise?

Ivan: Exercise can effectively mitigate the bone loss and muscle atrophy, but cannot completely stop them.

Kevin: What else do we have?

Ivan: There is a device called the Lower Body Negative Pressure(LBNP) trousers on the space station which can pull the body fluid to the lower body, thus reduce the effects of body fluid upshift. In Russia, that device is called Chibis garment. It can be used to countermeasure the cardiovascular deconditioning in space.

Kevin: What is a Penguin suit? I have heard that astronauts wear such suits in space.

Ivan: It is a suit with elastic strings inside which can exert a constant force on the person who wear it. When wearing the suit, the astronaut will constantly counteract the elastic force, quite like we counteract the gravity on the ground.

Kevin: That's all for the countermeasures?

Ivan: No. Nutrition is also an important part for the countermeasure.

Proper nutrition can help the metabolism of bones and muscles in space.

Kevin: Do we have a method to completely solve this problem?

Ivan: Artificial gravity can completely solve this problem.

Kevin: What is artificial gravity?

Ivan: By spinning a centrifuge or spinning the spacecraft as a whole, we can create artificial gravity. You may have seen the wheel like space facilities rotating slowly in science fiction, the science behind is artificial gravity. With artificial gravity we can solve micro-gravity related health problems completely. However, there are many technological challenges in creating artificial gravity in space. If we use a short arm centrifuge, the gravity gradient will be great and thus have adverse effects on the human body. If we adopt long arm centrifuge or spin the whole spacecraft in space, it will be technically very difficult.

Medical Monitoring and Medical Support

Kevin: How to see a doctor if an astronaut is sick in space?

Ivan: The astronauts are specially selected and trained, the chances of being sick in space is very small. Anyway, we have the medical monitoring and medical support team to help the astronauts.

Kevin: Is the health status of an astronaut constantly monitored in space?

Ivan: Yes! There are many medical sensors onboard, we can monitor the physiological parameters of the astronaut in a timely manner. The flight surgeon will have private conferences routinely with the astronaut onboard to make constant health evaluation.

Kevin: How to deal with a medical problem in space?

Ivan: The medial support team including the flight surgeon will evaluate the situation. If necessary, the astronaut can use the onboard medical equipment to perform medical checkup. The astronaut can call the flight surgeon for help or he can cope with it by himself according to the related procedures.

Kevin: It looks like seeing a doctor remotely.

Ivan: Yes. We call this practice telemedicine. Telemedicine enables preventive, diagnostic, and therapeutic care during months long flight in space, and ideally allows for seamless continuity of care before and after a mission.

Kevin: Do astronauts take training in medical care?

Ivan: Yes. Before launch, all astronauts are trained to use the medical assets that will be on board. Some astronauts will undergo 40 hours of paramedic-level training to qualify as crew medical officers. They become familiar with a checklist of foreseeable medical problems and emergency responses, such as a crewmate having difficulty breathing. They learn how to perform a periodic basic physical examination and how to handle the most common medical problems, such as space motion sickness, skin irritation, and back pain.

Kevin: What is a flight surgeon?

Ivan: Flight surgeons oversee the health care and medical training when astronauts are preparing for their mission and also take care of any medical issues that arise before, during, or after spaceflight.

Kevin: What certifications shall a flight surgeon have?

Ivan: Generally speaking, a flight surgeon is a physician who has specialized training in Aerospace Medicine. Flight surgeons must stay current on new technologies and advancements in telemedicine, pharmaceuticals, treatment protocols and diagnostic techniques to address the challenges of medical care for the crew in orbit.

Living and Working in Space

Kevin: What kinds of food do astronauts eat in space?

Ivan: Space food. The early space food looks like toothpaste and the taste is not good. But with the development of technology, the space foods nowadays are quite delicious and their types are abundant.

Kevin: What is the most typical space food?

Ivan: Dehydrated food is the most typical one. It is hydrated with water in space and can be warmed in the food warmer when eating them.

Kevin: What is the typical feature of the space food?

Ivan: They usually have long shelf life, because the food will be stored for a long time before being consumed in space.

Kevin: What about the dining table?

Ivan: The dining table in space is different from that on the ground. There is a magnet which can attract the forks and knives. Otherwise they may float away easily. There are also fastening taps and belts for fixing the foods.

Kevin: Can astronauts have fresh fruits in space?

Ivan: Yes, they can occasionally have fresh fruits when there is a newly arrived cargo resupply ship or when the new crew is arriving. The newly arrived astronauts can also take small amount of personal foods into space to share with their crewmates.

Kevin: Where do they live in space?

Ivan: The astronaut each has a small living room like a phonebooth. Inside the living room, there are a sleeping bag, a computer, and personal belongings.

Kevin: Why do they use the sleeping bag?

Ivan: The astronaut will sleep in the sleeping bag and the sleeping bag will be attached to the wall of the living room to avoid floating away during sleep.

Kevin: Can astronauts sleep well in space?

Ivan: The space station is rather noisy; the astronauts will use earplugs to shield the noise when living in space. Fortunately, the noise level in the sleeping quarter is usually smaller than the working area. Because the astronauts are very busy each day, most of them can have good sleep in space.

Kevin: How to go to the toilet in space?

Ivan: They can use the space toilets. The urine and the feces are separately collected. The urine will be collected through a tube-like device, preprocessed and then recycled to potable water, while the feces will be collected, sealed and disposed. There will be negative pressure in the collecting device to avoid floating away.

Kevin: What about the work and rest schedule of astronauts in space?

Ivan: It is quite similar to the ground schedule. They take a 24-hour work and rest schedule, although they experience a sunrise and a sunset every 90 minutes in space. That can help the astronauts to have the proper circadian rhythm. They will have weekends just like on the ground.

Kevin: What are the main jobs of astronauts in space?

Ivan: Their work mainly includes scientific experiments and the maintenance of the platform. In addition, taking care of themselves is also a routine job.

Kevin: What kind of experiments are there in space?

Ivan: The experiments in space mainly take advantage of the special space environment such as micro-gravity, vacuum, radiation, etc. Some are conducted inside the spacecraft, while others may be conducted outside. Take the ISS as an example, there have been thousands of experiments conducted.

Kevin: Do the experiments outside the space station require astronauts to take a spacewalk?

Ivan: Some experiments in space do need astronauts to take a spacewalk. Spacewalk is formally known as extravehicular activity, or shortly called EVA. EVA is very important for the construction and maintenance of large space structures such as the space station. It is one of the basic technologies of the space station program.

Kevin: How to conduct a spacewalk or EVA in space?

Ivan: First, you must have an EVA spacesuit. The EVA spacesuit is more like a small spacecraft which can support an astronaut to work in space up to 8 hours.

Kevin: Does the astronaut with the EVA spacesuit go out into space directly?

Ivan: No, the procedure is rather complex. The astronaut cannot go out immediately. The astronaut will stay in the airlock doing preoxygenation for some time. Then the sealed airlock will decrease the pressure gradually to zero. After that, the hatch to the space is opened and the astronaut can go out.

Kevin: What is preoxygenation and why?

Ivan: Preoxygenation means breathing oxygen for a period of time before EVA and thus replacing the nitrogen dissolved in our bodies. The pressure of the space station is 1 atmosphere which is similar to that on the ground, but the pressure of an EVA spacesuit is roughly 0.4 atmosphere. The pressure of the EVA spacesuit cannot be 1 atmosphere, otherwise the suit will be too stiff to operate. If we enter into the low-pressure EVA suit without doing preoxygenation, the resolved nitrogen will come out as bubbles in our bodies and the person will have decompression sickness. Only after preoxygenation, can the EVA astronaut perform EVA from the airlock.

Kevin: What is an airlock?

Ivan: An airlock is a small cabin between the space station module and the outside space. The airlock has two hatches, one is connected with the space station module, the other is connected to space. When the astronaut is in the airlock and fully prepared, the hatch connected with the space station will be closed. Then the airlock will depressurize to nearly zero. After that, the door connected to space is opened and the astronaut can go outside.

Kevin: How do the astronauts move around during EVA?

Ivan: The astronaut usually uses the foot restraints to fix the body and move with hands through handrails. They can also be easily moved to farther places by the robotic arm. During EVA, the astronaut will use tethers and hooks to avoid floating away.

Kevin: What types of EVA spacesuits do we have?

Ivan: EVA spacesuit can be further divided into orbital EVA spacesuit and planetary EVA spacesuit such as Lunar or Mars EVA spacesuits. During spaceflight, the astronaut will also have a pressure suit which is sometime called IVA spacesuit.

Kevin: What are the differences between the pressure suit and the EVA space suit?

Ivan: Pressure suit is used by astronauts during lift-off, critical operations in space such as rendezvous and docking, and landing. The purpose is to protect the astronaut from danger in case of emergency, for instance, depressurization of the spaceship. It has an umbilical cord connected to the mother ship to have power and oxygen supply.

Kevin: What about the EVA space suit?

Ivan: The EVA spacesuit is more like a small spacecraft. The portable life support system in the backpack can create a habitable environment. It can maintain normal air pressure, oxygen partial pressure, humidity and temperature inside, the carbon dioxide can be removed.

Kevin: Can you briefly introduce the structure of an EVA spacesuit?

Ivan: The EVA spacesuit has a helmet on the top which is mounted with sunshield and illumination devices. The helmet is fixed and the astronaut can turn the head inside to observe different places. The torso is hard and the limbs are flexible. The portable life support system is mounted in the backpack. The EVA spacesuit have several layers including the restrain layer, the air-tight layer, the thermal insulation layer, the micrometeorite layer, and outer cover, etc. Control of the suit is via a panel on the chest, with the markings in mirror image. The cosmonaut views the panel using a mirror on the wrist of the suit.

Kevin: Which country or countries can produce EVA spacesuits?

Ivan: Russia, the United States and China are the only three nations that have successfully developed EVA spacesuits. The Russian EVA spacesuit is called Orlan which has been used for decades. The Orlan spacesuit is quite robust and requires shorter pre-breathing time, about 30 minutes.

The U.S. EVA spacesuit is called Extravehicular Mobility Unit (EMU), which is different in configuration with Orlan. It consists of the upper part and the lower part and the connection is in the waist. In comparison, the Orlan is opened from the backpack hatch and the astronaut can go inside from the hatch. The EVA spacesuit of China is called Feitian which is more similar to Orlan.

Kevin: You have mentioned that robotic arm can be used to transport the EVA astronauts, can you introduce the robotic arm?

Ivan: Robotic arm is also called the space manipulator. Robotic arm is a very important instrument for the construction and operation of a space station. It can be used to transport cargo or crew on the space station and even some cargo spaceships need to be captured or re-berthed by the robotic arm.

Kevin: How to operate the space manipulator?

Ivan: It is very difficult to operate a space manipulator in space. The astronaut will take extensive training on the ground. Take the ISS as an example, the astronaut will control the Canadarm2 robotic arm via control sticks in the Copula Module where they can observe the operation directly.

Kevin: Can you introduce the robotic arms on ISS?

Ivan: The most famous robotic arm on ISS is Candarm2 which is 57.7 feet long and is equipped with Dextre robotic hand. It has been used to capture and dock a number of unpiloted spacecrafts that deliver science payloads and cargo to the space station. The Japanese Module Kibo is also equipped with a robotic arm called Japanese Experiment Module

Remote Manipulator System (JEMRMS). It is intended for supporting experiments to be conducted on Kibo's Exposed Facility (EF) or for supporting Kibo's maintenance tasks.

Kevin: Thank you for the introduction!

Ivan: You are welcome!

Section 3. Astronaut Landing

场景介绍：航天员返回地球的过程同样充满了挑战和风险，每个环节都要做到精准无误，离轨、再入、黑障、开降落伞、反推发动机点火、着陆、切伞……

Kevin: How long can an astronaut stay in space?

Ivan: It depends on the mission duration. For short duration flight, it may last only a few days. For long duration flight, it may last 6 months. The record of the longest single flight in space is 438 days.

Kevin: Why cannot astronauts live in space for longer period time?

Ivan: Because micro-gravity environment has adverse effects on human body, the current technologies and countermeasures can only partially alleviate the changes of human body. Changes of human body within six months flight in space are proved to be reversible, that is the reason why the International Space Station expedition crew rotate about every 6 months.

Kevin: What preparations do they make when returning to the Earth?

Ivan: There are a lot of work to do. Physically, the astronauts will supplement body fluid upon return. That will help them to re-adapt to the 1G environment upon landing. Because the astronauts lose some body fluid in micro-gravity.

Kevin: What is the procedure of landing?

Ivan: Take the spaceship returning to the Earth from the space station as an example. First, the spaceship will detach the space station and retreat to a safe distance with the space station. Then there will be de-orbit burns of the thrusters of the service module. When decreased to certain altitude, the service module will be jettisoned and the return module enters into the reentry corridor. At certain height in the atmosphere, the pilot and main parachutes will be deployed and upon landing the cushioning airbags or the landing thrusters will make the landing less stressful. That is just a rough description.

Kevin: Where will the spaceship land?

Ivan: It depends. For the normal landing, the spaceship will land on the landing site. The landing site of Shenzhou spaceship of China is in the grassland of inner Mongolia, the Russian Soyuz will land in Kazahkstan, while the American Crew Dragon will make a splash down in the Atlantic Ocean. For emergency landing, it can be anywhere in the world, forest, desert, grassland, water......

Kevin: How to recover and rescue the astronauts after landing?

Ivan: The tracking system will make constant prediction and tracking of the landing position. The rescue team is fully prepared to recover the astronauts. For the ocean splash down, marine vessels equipped with helicopters will be used for the rescue. For the landing on the ground, helicopters and rescue trucks and cars will be used.

Kevin: I have seen live TV broadcast that the landed astronauts lie in the chair in front of the return capsule, why?

Ivan: Long duration flight in micro-gravity will weaken the muscles and bones of astronauts, though they have taken extensive daily exercises in space. Upon return to 1G environment, they must take time to re-adapt. If they stand up too early, they may faint and collapse due to orthostatic intolerance.

Kevin: How long will it take for an astronaut to completely recover after a long duration flight?

Ivan: It roughly takes about 45 days for the astronaut to reach or exceed the preflight test values. Some crewmembers subjectively indicated that they need for a longer rehabilitation period.

Kevin: What is rehabilitation?

Ivan: Rehabilitation means the recovery of astronauts after space flight.

Kevin: How to implement the rehabilitation?

Ivan: The rehabilitation programs are tailored specifically for each individual according to his on her test result. The 45 day rehabilitation program can be divided into 3 phases. Phase 1 starts on the landing day and places emphasis on ambulation, flexibility, and muscle strengthening. Phase 2 adds proprioceptive exercise and cardiovascular conditioning. Phase 3 (the longest phase) focuses on functional development.

Kevin: Thank you for the explanation!

Ivan: You are welcome!

Section 4. Terminology and Abbreviations

induced environment　诱导环境

zero-G plane　失重飞机

parabolic flight　抛物线飞行

drop tower　落塔

ionizing radiation　电离辐射

accumulated exposure dose　累积照射剂量

communication delay　通讯延迟

in-situ resource utilization technology　原位资源利用技术

space motion sickness　空间运动病

bone loss　骨质流失

muscle atrophy　肌肉萎缩

hypothesis　假说

sensory conflict theory　感觉冲突理论

osteoporosis　骨质疏松症

space medicine　航天医学

vibration free treadmill　隔振跑台

bicycle ergometer　自行车功量计

Lower Body Negative Pressure　下体负压

chibis garment　气比斯服

cardiovascular deconditioning　（长期失重后）心血管失适应

Penguin suit　企鹅服

artificial gravity　人工重力

medical monitoring and medical support　医学监督与医学保障

flight surgeon　航天医生

telemedicine　遥医学

aerospace medicine　航空航天医学

shelf life　保质期

circadian rhythm　昼夜节律

preoxygenation　预吸氧

decompression sickness　减压病

airlock　气闸舱

foot restraint　脚限制器

handrail　扶手

tether　系绳

hook　挂钩

portable life support system　便携式生命保障系统

oxygen partial pressure　氧分压

restrain layer　限制层

air-tight layer　气密层

Orlan　奥兰舱外航天服

Extravehicular Mobility Unit (EMU)　舱外机动装置

Copula Module　穹顶舱

Japanese Experiment Module Remote Manipulator System (JEMRMS)
日本实验舱遥操作系统

Exposed Facility (EF)　暴露设施

service module　服务舱

splash down　水上溅落

orthostatic intolerance　立位耐力不良

rehabilitation period　后恢复期

EF: Exposed Facility　暴露设施

EMU: Extravehicular Mobility Unit　舱外机动装置

JEMRMS: Japanese Experiment Module Remote Manipulator System
日本实验舱遥操作系统

LBNP: Lower Body Negative Pressure　下体负压

Section 5. Extended Reading Material

1. Exercising in space prevents astronauts from fainting when returning to the Earth

Astronauts who exercised regularly during their missions and received intravenous saline fluids upon returning to the Earth did not experience orthostatic hypotension, the drop in blood pressure caused by blood flowing toward the feet from the brain, according to researchers at the University of Texas Southwestern Medical Center.

Orthostatic hypotension typically occurs when a person stands up after sitting or lying down for an extended period of time - known colloquially as a "head rush" - but returning to the Earth from the zero-gravity environment of space can also trigger the condition, which can cause dizziness and fainting.

"One of the biggest problems since the inception of the manned space program has been that astronauts have fainted when they came down to the Earth. The longer the time spent in a gravity-free environment in space, the greater the risk appeared to be," Dr. Benjamin Levine, the study's senior author and a professor of exercise sciences at UT Southwestern Medical Center, told media outlets.

"This problem has bedeviled the space program for a long time, but this condition is something ordinary people often experience as well."

In the study, 12 astronauts (eight men and four women) between the ages of 43 and 56 did endurance and resistance exercise training up to two hours every day for six months while in space. When they returned to the Earth, they received a saline infusion.

The astronauts' blood pressures were also taken on each heartbeat before embarking to space, in space and after returning to the Earth. The measurements showed very little change in blood pressure during each phase. In addition, the astronauts did not experience any fainting or dizziness after coming back to the Earth.

"What surprised me the most was how well the astronauts did after spending six months in space. I thought there would be frequent episodes of fainting when they returned to the Earth, but they didn't have any. It's compelling evidence of the effectiveness of the countermeasures - the exercise regimen and fluid replenishment," Levine said.

Vocabulary

regularly：定期地、有规律地

intravenous：静脉的

saline fluids：生理盐水

orthostatic hypotension：立位性低血压

colloquially：通俗地、口语地

head rush：头晕

trigger：出发

dizziness：头晕

fainting：昏迷

inception：起初

gravity-free environment：无重力环境

senior author：资深作者

bedevil：使痛苦

endurance and resistance exercise：耐力和抗阻锻炼

saline infusion：生理盐水输液

compelling evidence：令人信服的证据

countermeasures：对抗措施

regimen：方案

replenishment：补充

2. A very good start

The first spacewalk to service the Alpha Magnetic Spectrometer (AMS) could not have gone better. Lead spacewalker ESA astronaut Luca Parmitano is imaged here hitching a ride on the International Space Station's 16-metre long robotic arm to kick 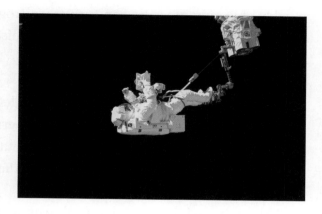 off the first of four ventures to service the particle physics detector on 15 November, 2019.

While all spacewalks are a carefully planned and detailed affair, the four spacewalks for AMS are exceptionally difficult as the bus-sized dark matter detector was never designed to be maintained in space. But after three successful years of delivering ground breaking science, the decision was made to extend its lifetime.

The cooling pumps for AMS-02 need maintenance and without them it will no longer be able to collect data on the cosmic rays that are bombarding our planet. The first question spacewalk designers had to answer was whether this was even possible.

The first spacewalk proved it was not only possible, but thanks to the planning and training that began as early as 2017, Luca and his spacewalking partner Andrew Morgan could achieve more than scheduled – setting them in good stead for the next phase.

The spacewalk began, as they all do, with "prebreathing" for up to two hours. Similar to scuba divers, astronauts can suffer from the 'bends': quickly changing pressure can turn the nitrogen in human bodies into bubbles with serious symptoms. To avoid this, astronauts breathe pure oxygen to purge their bodies of nitrogen.

Luca and NASA astronaut Drew Morgan left the depressurised Quest airlock at 13:10 CET (12:10 GMT), with Luca grabbing the ride to AMS on the robotic arm controlled by NASA astronaut Jessica Meir while Drew ferried handrails and equipment by hand to the worksite.

The main task of this spacewalk was to remove the debris shield covering AMS, with an estimated three hours portioned for this task. Luca and Drew managed to jettison the debris shield to burn up safely in the Earth's atmosphere well ahead of schedule.

Luca and Drew also installed three handrails in the vicinity of AMS to prepare for the next spacewalks and removed zip ties on the AMS' vertical support strut.

Amazingly, the duo were still well ahead of the six hours planned for the main task of removing the debris shield.

When time permits, mission control give spacewalkers some "get ahead" tasks. Although there were no get-ahead tasks planned for this spacewalk, the duo was so far ahead of schedule that mission control agreed they could continue the work originally planned for the second AMS spacewalk. Luca removed the screws from a carbon-fibre cover under the insulation and passed the cover to Drew to jettison once again.

The pair cleaned up, took some photos of their killer views, gathered tools, and made their way back to the airlock, clocking in 6 hours and 39 minutes for this promising start to AMS maintenance.

Vocabulary

Alpha Magnetic Spectrometer (AMS)：阿尔法磁谱仪

hitching：搭便车

kick off：开始

extend its lifetime：延长它的寿命

spacewalk designer：太空漫步设计者

pre-breathing：预吸氧

scuba diver：水肺潜水员

bends：减压病

symptom：症状

purge：清除

depressurised：减压的

Quest airlock：探索号气闸舱

ferry：运送

handrail：扶手

debris shield：空间碎片防护板

jettison：抛弃

vicinity：附近

zip tie：扎线带

support strut：支撑杆

screw：螺丝

carbon-fibre：碳纤维

insulation：绝缘、隔离

killer：断路器

promising：有希望的

3. Russian Humanoid Robot Fedor Announces Full Implementation of Flight Test Program

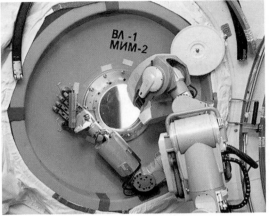

Fedor, Russia's first robot to fly to the International Space Station (ISS), returned to the Earth on board the Soyuz MS-14 spacecraft, space agency Roscosmos said in the early hours of Saturday.

Fedor, which landed in Kazakhstan, 147 km south-east of Zhezkazgan, said that the flight test program was fully complete. The corresponding entry appeared on his Twitter page.

"Landed. The test of the new landing system passed without remarks. The test program has been fully completed. The lander has been taken under protection," the robot wrote.

"Fedor" also noted that the descent from orbit was normal.

The spacecraft loaded with Fedor was launched into space on 22 August and docked to the ISS five days later. On Friday, the vehicle undocked and was de-orbited.

"The SoyuzMS14 spacecraft descent module carrying SkybotF850 humanoid robot successfully landed. All the reentry and landing operations went as expected!" Roscosmos wrote on Twitter.

The Soyuz MS-14 spacecraft with Fedor on board was launched on 22 August, and five days later, the spacecraft docked to the ISS after a second attempt.

Vocabulary

entry：进入

descent：下降

load：装载

dock：对接

undock：脱离

descent module：降落舱

humanoid：类人的

reentry：再入

4. Record-Setting NASA Astronaut, Crewmates Return from Space Station

After setting a record for the longest single spaceflight in history by a woman, NASA astronaut Christina Koch returned to the Earth Thursday, along with Soyuz Commander Alexander Skvortsov of the Russian space agency Roscosmos and Luca Parmitano of ESA (European Space Agency).

The trio departed the International Space Station at 12:50 a.m. EST and made a safe, parachute-assisted landing at 4:12 a.m. (3:12 p.m. Kazakhstan time) southeast of the remote town of Dzhezkazgan, Kazakhstan.

Koch's extended mission will provide researchers the opportunity to observe effects of long-duration spaceflight on a woman as the agency plans to return humans to the Moon under the Artemis program and prepare for human exploration of Mars.

Koch was launched March 14, 2019, alongside fellow NASA astronaut Nick Hague and Russian cosmonaut Alexey Ovchinin. Her first journey into space of 328 days is the second-longest single spaceflight by a U.S. astronaut and also places her seventh on the list of cumulative time in space for American astronauts with one or more missions.

Supporting NASA's goals for future human landings on the Moon, Koch completed 5,248 orbits of the Earth and a journey of 139 million miles, roughly the equivalent of 291 trips to the Moon and back. She conducted six spacewalks during 11 months on orbit, including the first three all-woman spacewalks, spending 42 hours and 15 minutes outside the station. She witnessed the arrival of a dozen visiting spacecrafts and the departure of another dozen.

Following post-landing medical checks, the crew will return to the recovery staging city in Karaganda, Kazakhstan, aboard Russian helicopters. Koch and Parmitano will board a NASA plane bound for Cologne, Germany, where Parmitano will be greeted by ESA officials for his returning home. Koch will continue home to Houston. Skvortsov will board a Gagarin Cosmonaut Training Center aircraft to return to his home in Star City, Russia.

The Expedition 61 crew contributed to hundreds of experiments in biology, Earth science, human research, physical sciences and technology development, including improvements to the Alpha Magnetic Spectrometer in an effort to extend its life and support its mission of looking for evidence of dark matter and testing 3D biological printers to print organ-like tissues in micro-gravity.

Vocabulary

Soyuz Commander：联盟飞船指令长

Russian space agency：俄罗斯航天局

trio：三人

depart：离开

parachute-assisted landing：降落伞辅助下的着陆

long-duration spaceflight：长期飞行

fellow NASA astronaut：NASA 航天员同事

on orbit：在轨

departure：离开

post-landing medical checks：飞行后医学检查

recovery：回收

board：登机 / 船

Gagarin Cosmonaut Training Center：加加林航天员训练中心

Star City：星城

Alpha Magnetic Spectrometer：阿尔法磁谱仪

dark matter：暗物质

organ-like tissues：器官类组织

5. NASA Selects Proposals to Advance Understanding of Space Weather

NASA has selected three proposals for concept studies of missions that could help us better understand the dynamic space weather system driven by the Sun that manifests near Earth. The proposals examine what drives different parts of that system and ultimately could help us predict and mitigate its effects on spacecrafts and astronauts, as NASA's Artemis program looks to send the first woman and the next man to the Moon by 2024.

"NASA's research to understand the space we travel through relies on exploring key details about a vast system from the Sun, to the Earth, to the edges of the solar system," said Peg Luce, deputy director for heliophysics in the Science Mission Directorate at NASA Headquarters in Washington. "Each of these proposals could add a significant tool from a unique vantage point to help us understand that system."

Each of these Heliophysics Mission of Opportunity proposals will receive $400,000 to conduct a nine-month mission concept study. After the study period, NASA will choose one proposal to go forward to launch. Each potential mission has a separate launch opportunity and time frame.

The proposals were selected based on potential science value and feasibility of development plans. The total cost for the mission ultimately chosen will be capped at $55 million and is funded by NASA's

Heliophysics Explorers' program.

The selected proposals are:

Extreme Ultraviolet High-Throughput Spectroscopic Telescope (EUVST) Epsilon Mission

Aeronomy at Earth: Tools for Heliophysics Exploration and Research (AETHER)

Electrojet Zeeman Imaging Explorer (EZIE)

Vocabulary

concept study：概念研究

manifest：出现

ultimately：最终

mitigate：减轻

heliophysics：太阳物理学

launch opportunity：发射机会

time frame：时间范围

feasibility：可行性

ultraviolet：紫外的

spectroscopic：光谱学的

aeronomy：高层大气物理学

electrojet：电喷流

在奋进号航天飞机中舱穿 EMU 舱外航天服
Donning Extravehicular Mobility Unit（EMU）on the Mid Deck
of the Space Shuttle "Endeavour"

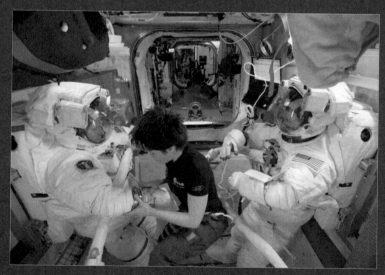

气闸舱内准备穿美国舱外航天服出舱
Donning EMU Space Suit of the U.S. in the Airlock

机械臂帮助下的太空行走
EVA Helped by Robotic Arm

不系绳的太空行走 Untethered EVA

加拿大机械臂 Candada Arm-2

操作机械臂 Robotic Arm Operation

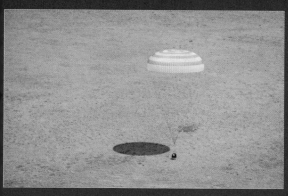

飞船降落 Spaceship Landing

搜救航天员 Astronaut Rescue

杨利伟圆满完成神五任务凯旋
Yang Liwei Returned to Earth Safely

航天员返回后座在椅子上
Resting on Chair after Landing

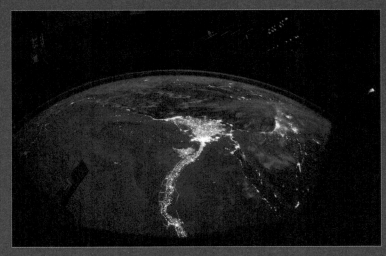

美丽的地球夜景 Night of Earth from Space

月球上看地球升起 Earth Rise Seen from Moon

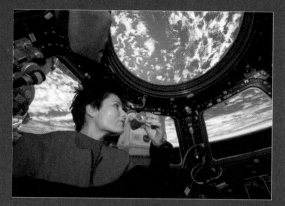

喝咖啡 Drinking Coffee

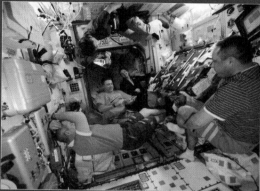

太空吃饭 Eating in Space

国际空间站卧室 Sleeping Room on ISS

航天飞机卧室 Living Quarter in Space Shuttle

剪发 Haircutting in Space

太空搬运 Cargo Transportation in Space

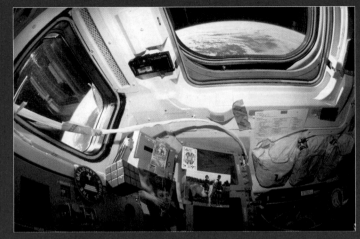

太空中的个人物品
Personal Belongings in Space

太空中欣赏风景
Enjoying Beautiful Views
in Space

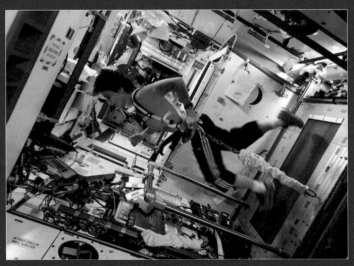

太空跑步机
Space Tredmill

太空做饭
Preparing Food in Space

太空中的新鲜水果
Fresh Fruits in Space

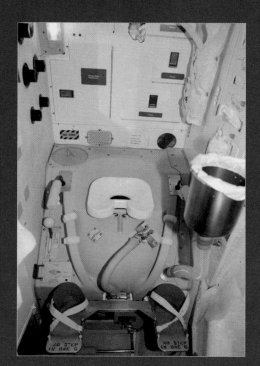

太空卫生间 Space Toilet

太空自行车 Bicycle Ergometer

陈冬准备进餐
Preparing Food by Taikonaut Chen Dong

神九乘组太空合照
Group Photo of Shenzhou-9 Crew

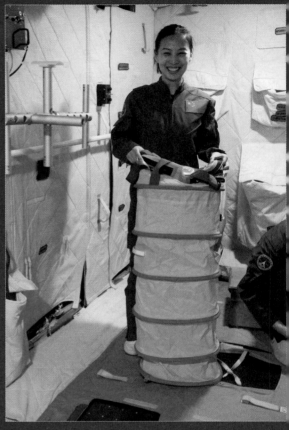

下体负压装置
Lower Body Negative Pressure Device

中国首飞航天员杨利伟太空吃东西 Eating Food by First Taikonaut Yang Liwei

Chapter 5
Robotic Explorations

　　航天任务中很多是不载人的，这些无人探索任务深刻地改变着人类社会的生产和生活方式，促进了地球科学、天文学等科学领域的发展，如全球导航卫星系统，气象卫星、海洋卫星、大气卫星等地球观测卫星，以及太空望远镜等；此外，人类还将无人探测器和漫游车发射到了月球和火星，并利用各种探测器对太阳系内的天体如太阳、金星、水星、木星、土星等进行了探索研究，甚至开展了小行星采样返回任务。

Section 1. Navigation Satellites

　　场景介绍：导航定位与我们的日常生活息息相关，航天专家将带我们了解不同的卫星定位导航系统：美国的 GPS，中国的北斗，俄罗斯的格洛纳斯，欧洲的伽利略……

Kevin: Space exploration is expensive. Can you introduce some of the benefits of space exploration to humanity?

Ivan: Of course. Space exploration is really costly, but the benefits are also enormous. GNSS, or Global Navigation Satellite System, is just one example.

Kevin: Is GPS a kind of Global Navigation Satellite System?

Ivan: Yes. GPS stands for Global Positioning System and it is the best known of these systems. GNSS has a wide range of applications in our daily life and is becoming a booming business. It brings countless job positions and trillions of dollars of revenue for the world.

Kevin: What is the principle behind the satellite navigation?

Ivan: Generally speaking, the navigation satellite can provide positioning, navigation and timing services. The principle is that: at any place on the Earth, the receiver can receive signals from at least four satellites simultaneously. Once a device knows its distance from at least four satellites, it can use geometry to determine its location on Earth in three dimensions.

Kevin: How many navigation satellite systems do we have in the world?

Ivan: There are four systems in the world, the GPS of the Unites States, the GLONASS of Russia, the Galileo of the European Union and the Beidou of China.

Kevin: Can you tell me the components of a GNSS?

Ivan: The system usually consists of three segments: the space segment, the control segment and the user segment. Take GPS as an example, the space segment consists of 24 operating satellites that transmit one-way signals that give the current GPS satellite position and time. The control segment consists of worldwide monitor and control stations that maintain the satellites in their proper orbits and adjust the satellite clocks. While the user segment consists of the receiving equipments which receive the satellite signals and calculate the position.

Kevin: When were these systems built?

Ivan: GPS is a military service operated by the U.S. Air Force. The first satellite of GPS was launched in 1978 and the constellation was completed in 1993. GPS has been upgraded continuously and now is the third generation. The GLONASS system started satellite deployment in 1982 and the constellation was completed in 1995. The first satellite of the Galileo system was launched in 2011 and completed in 2018. The Beidou system of China includes Beidou 1, Beidou 2 and Beidou 3. Beidou 1 system has retired. Beidou 3 started construction in 2009 and the last satellite expected to be launched in June 2020 is delayed due to technical problems, but it does not affect the operation of Beidou.

Kevin: How many satellites are there in each system?

Ivan: GPS is a constellation of 21 operational satellites and 3 on-orbit spares flying 20,350 km above the surface of the Earth. Similar to GPS, the fully operational GLONASS constellation also consists of 24 satellites, with 21 used for transmitting signals and three for in-orbit spares. The fully deployed Galileo system consists of 30 satellites including 24 operational satellites and 6 active in-orbit spares. While the Beidou system of China now has 39 satellites in orbit.

Kevin: What about the navigation accuracy?

Ivan: A GNSS usually provides two different navigation accuracies, one is for the general public, the other is for the military or for the restricted users. For the public, the accuracy may be as big as 10 meters, while for the military, the accuracy may be as small as one centimeter.

Kevin: Thank you for the introduction!

Ivan: You are welcome!

Section 2. Earth Observation

场景介绍：地球观测卫星可以帮助我们更好地了解地球，我们将逐一认识不同种类的地球观测卫星：气象卫星、海洋卫星、大气卫星、侦察卫星……

Kevin: In addition to the navigation satellites, what other types of satellites do we have?

Ivan: There are many different types of satellites operating in Earth orbit. The satellites can be categorized in different ways. For instance, they can be categorized according to their orbits as polar orbit satellites, equator orbit satellites, geostationary satellites, etc. or according to the functions such as ocean observing satellites, land observing satellites, meterological satellites, communication satellites, reconnaissance satellites and so on.

Kevin: What is a reconnaissance satellite?

Ivan: A reconnaissance satellite, or a spy satellite, is an Earth observation satellite or communication satellite deployed for military or intelligence applications.

Kevin: What can a reconnaissance satellite do?

Ivan: As a military satellite, it may carry out photographic surveillance, gather electronic intelligence, detect nuclear explosion and provide early warning of strategic-missile launchings.

Kevin: What is a meterological satellite?

Ivan: A meterologcial satellite, also called a weather satellite, can give meteorologists a view of weather patterns over a very large area, thus enables them to track large weather patterns and make more accurate predictions of future weather behavior.

Kevin: What sensors are there on the weather satellite?

Ivan: There are many sensors. In addition to visual monitoring, weather satellites may also use thermal imaging, X-ray sensors, high-energy proton detectors, alpha detectors, and energetic particle sensors.

Kevin: What orbits do the weather satellites usually take?

Ivan: They travel in either geostationary or polar orbits. Geostationary orbits provide views of the same portion of the Earth's surface 24 hours a day. While polar orbits can provide a view of the entire planet over the course of a 12 hour orbit.

Kevin: What are the advantages of studying or observing the Earth from space with satellites?

Ivan: Comparing with the traditional methods such as by airplanes, the satellite-based observation is highly efficient and cost-effective. So, there are a large number of Earth observation satellites in orbit such as the land satellites, the atmosphere satellites, the ocean satellites, etc.

Kevin: Can you explain these Earth observation satellites?

Ivan: Ok. Land satellites focus on the land of our Earth. Take the U.S. Landsat program an example, it collects data on the forests, farms, urban areas and freshwater of our planet. These data can help us to better understand the environmental changes, manage agricultural practices, allocate scarce water resources, respond to natural disasters and more. It not only collects images of the Earth's landscapes in visible, near-infrared and shortwave infrared light, but also measures the temperature of land surfaces.

Kevin: Can you introduce other Earth observation satellites?

Ivan: Yes. There are many Earth observation satellites focusing on different aspects of the Earth, for instance, the ocean satellite programs of China, the Sentinel Missions of European Copernicus Program, NASA's Aura and OCO-2 satellites, JAXA's GOSAT and ALOS-2 satellites, etc. These are just a few examples. The main differences of these satellites are their orbits and the sensors they carry.

Kevin: I have read that some nanosats or CubeSats can also observe the Earth?

Ivan: Yes. A nanosatellite is a type of small satellite with mass from 1 kg to 10 kg. A CubeSat is a type of nanosatellite defined by the CubeSat Design Specification (CSD), unofficially called the CubeSat standard. The dimensions of 1U CubeSat is 10 cm × 10 cm × 11.35 cm, while 2U CubeSat is 10 cm × 10 cm × 22.70 cm. Many nanosats are designed to observe the Earth.

Kevin: How many nanosats have been launched?

Ivan: Nanosat is cheap and it develops very quickly. Up to now, 1317 nanosats have been launched.

Kevin: Thank you for the introduction!

Ivan: You are welcome!

Section 3. Space Observatory

场景介绍：太空望远镜要比地面上的天文望远镜具备更大的优势，它能有效避免大气对观测的影响。航天专家向凯文介绍了著名的太空望远镜：哈勃太空望远镜、詹姆斯·韦伯太空望远镜、钱德拉 X 射线天文台……

Ivan: Among the satellites in Earth orbit, some are "looking down" like the Earth observation satellites, some are "looking up" like the space observatories.

Kevin: What is a space observatory?

Ivan: A space observatory is a space telescope observing the universe.

Kevin: Why do we need a space telescope?

Ivan: About 400 years ago, human observed the universe with the naked eye. Since Galileo used telescopes to see the sky in 1610, telescopes have grown in size, complexity and power, but they cannot escape the disturbance of the atmosphere during observation. In space, there will be no distortion of the atmosphere and we can have an unobstructed view of the universe. That is the advantage of the space telescope.

Kevin: Can you introduce the famous space telescopes in the world?

Ivan: Hubble Space Telescope (HST), the Spitzer Space Telescope, the Chandra X-Ray Observatory (CXO), the XMM-Newton, the Compton

Gamma-Ray Observatory (CGRO) and the James Webb Space Telescope to be launched.

Kevin: Can you introduce them in detail?

Ivan: The Hubble Space Telescope is the most famous one. It is the first major optical telescope in space which has been used by scientists to observe the most distant stars and galaxies as well as the planets in our solar system. Hubble's launch and deployment in April 1990 marked the most significant advance in astronomy since Galileo's telescope.

Kevin: Can you briefly introduce the achievements of the Hubble Space Telescope?

Ivan: Hubble has exceeded the designed goal, operating and observing the universe for almost 30 years. During its time in orbit, the telescope has taken more than 1.4 million observations, and astronomers have used that data to publish more than 17,000 peer-reviewed scientific publications on a broad range of topics.

Kevin: What are the differences of other space telescopes comparing with the Hubble Space Telescope?

Ivan: You know each space-based observatory detects different spectrum of radiation in the universe. The Hubble Space Telescope as an optical telescope mainly detects visible light.

Kevin: What radiation does the Spitzer Space Telescope detect?

Ivan: Spitzer is designed to detect infrared radiation, which is primarily heat radiation.

Kevin: What does the infrared radiation detection means to us?

Ivan: That means Spitzer can peer into cosmic regions that are hidden from optical telescopes, including dusty stellar nurseries, the centers of galaxies, and newly formed planetary systems. Spitzer's infrared eyes also allows astronomers see cooler objects in space, like failed stars (brown dwarfs), extrasolar planets, giant molecular clouds, and organic molecules that may hold the secret to life on other planets.

Kevin: What about the X-ray space telescopes?

Ivan: The Chandra X-Ray Observatory is a NASA telescope that looks at black holes, quasars, supernovas, and the like – all sources of high energy in the universe. It shows a side of the cosmos that is invisible to the human eye.

Kevin: What about XMM-Newton?

Ivan: It can also take high-resolution images and spectra of X-ray sources of the universe. The XMM-Newton has been able to measure for the first time the influence of the gravitational field of a neutron star on the light it emits. This measurement provides much better insight into these objects.

Kevin: Can you introduce the future James Webb Space Telescope?

Ivan: It was formerly known as the "Next Generation Space Telescope" (NGST) and renamed in Sept. 2002 after a former NASA administrator, James Webb. It will be a large infrared telescope with a 6.5-meter primary mirror.

Kevin: When will the telescope be launched?

Ivan: It will be launched on an Ariane 5 rocket from French Guiana in 2021.

Kevin: Thank you for the introduction!

Ivan: You are welcome!

Section 4. Probe to Celestial Bodies

　　场景介绍：地外着陆探测器为载人探索打下坚实基础，人类已经将着陆探测器降落到了月球、火星等地外天体。就让航天专家带我们深入了解一下这些探测器吧……

Kevin: Have we landed robotic rovers to other planets?

Ivan: Yes. Although our human beings have set our footprints only on the Moon, many unmanned spacecraft have been launched to explore different celestial bodies in our solar system.

Kevin: I have heard that if we want to launch a mission to another planet, for instance, the Mars, we cannot launch at any time we want and there is a launch window. What is a launch window?

Ivan: Actually, the launch window is a period of time. To make sure a spacecraft and a planet such as the Mars arrive at the same place at the same time, the spacecraft must launch within a particular window of time. This window is called the launch window and, depending on the target, it can be a few minutes or as much as a few weeks in length. If a spacecraft is launched too early or too late, it will arrive in the planet's orbit when the planet is not there. For Mars mission, there is only one launch window every 26 months.

Kevin: I have heard that design the trajectory of a spacecraft is very complicated. Can you briefly explain that?

Ivan: The design is complex. But the basis is still the Newton's laws of motion in classical physics. Hohmann transfer orbit is frequently used in the trajectory design. The Hohmann transfer orbit is an elliptical orbit used to transfer a spacecraft between two circular orbits of different altitudes, in the same plane. Sometimes we also use the gravity of a planet flyby to accelerate the spacecraft in deep space exploration. These are some basic knowledges of the orbit design.

Kevin: When was the first spacecraft soft-landed on the moon?

Ivan: The moon is the closet celestial body to the Earth. The Soviets made the first moon fly with Luna 1 in 1959, photographed the far side of the moon with Luna 3 in the same year, and made the first soft-landing of unmanned Luna 9 spacecraft in 1966. The first soft-landing of the United States is the Surveyor 1 in the same year.

Kevin: What about China?

Ivan: Chang'e 3 of China made a soft-landing on the moon in 2013, and Yutu(Jade Rabbit) became the first lunar Rover of China. Chang'e 4 made a soft-landing on the far side of the moon for the first time in human history. The name of the rover in Chang'e 4 is called Yutu 2(Jade Rabbit 2) which is still in operation.

Kevin: What is the future plan of the moon exploration?

Ivan: China is planning to send Chang'e 5 to the moon in 2020 and lunar samples will be collected and returned. This will be the last step of China's Lunar Mission. Returning to the moon (Manned lunar mission) is now a hot spot.

Kevin: What about the robotic Mars soft-landing missions?

Ivan: Landing on the Mars is extremely difficult and that is why the Mars is dubbed as the tomb of spacecraft. The first soft-landing on the Mars was the Viking 1 in 1976. After that a number of Mars rovers have landed and operated on the Mars surface such as the Phoenix Mars Lander, Spirit Mars rover, Opportunity Mars rover, and some of them are still working such as the Curiosity rover and InSight Mars Lander.

Kevin: What are the purposes of the Mars explorations?

Ivan: The purposes mainly include four aspects: determine if life ever arose on the Mars, characterize the climate of the Mars, characterize the geology of the Mars, and prepare for the human exploration of the Mars.

Kevin: What robotic missions will be launched in the near future?

Ivan: The United States is going to launch a new mission: the Mars Perseverance Rover and China is going to launch the Tianwen Mars Mission. Both of the missions are planned to be launched in July 2020.

Kevin: What other celestial bodies have our humans landed?

Ivan: Asteroids. Exploration of asteroids can help scientists to study the origin of our universe. Many asteroids have been explored by unmanned spacecrafts and even samples of asteroids have been collected and returned by missions such as the Hayabusa 1 and 2 mission of Japan.

Kevin: What other places in the universe have human explored?

Ivan: Except the moon and the Mars, humans have sent robotic spacecraft to study a number of different celestial bodies in the solar system, for instance, the Juno for Jupiter, Parker Solar Probe to study the Sun, the New Horizons to study Pluto and the BepiColombo to Mercury, etc. But the spacecraft have not soft-landed on those celestial bodies.

Kevin: What is the farthest place a spacecraft has ever reached in the universe?

Ivan: Voyager 1 which was launched in 1977 is now 22.25 billion km from the Earth. Its primary destination is Jupiter and Saturn, it now reaches the edge of our solar system.

Kevin: Thank you for the introduction!

Ivan: You are welcome!

Section 5. *Terminology and Abbreviations*

Global Navigation Satellite System　全球导航卫星系统

Global Positioning System　全球定位系统

GLONASS　格洛纳斯（俄罗斯导航卫星系统）

Galileo　伽利略（导航卫星系统）

Beidou　北斗（导航卫星系统）

polar orbit satellite　极轨卫星

meterological satellite　气象卫星

reconnaissance satellite　侦察卫星

Geostationary orbits　地球同步轨道

Landsat　地球资源卫星

Sentinel mission　哨兵任务

CubeSats　立方体卫星

Nanosat　微型卫星

CubeSat Design Specification (CSD)　立方体卫星设计规范

space observatory　空间天文台

Spitzer Space Telescope(SST)　斯皮策太空望远镜

Chandra X-Ray Observatory (CXO)　钱德拉 X 射线天文台

Compton Gamma-Ray Observatory (CGRO)　康普顿伽马射线天文台

Quasar　类星体

Supernova　超新星

launch window　发射窗口

Hohmann Transfer orbit　霍曼转移轨道

Viking　海盗号

Phoenix　凤凰号

Spirit　勇气号

Opportunity　机遇号

Curiosity　好奇号

InSight　洞察号

Perseverance　毅力号

Hayabusa 2　隼鸟 2 号

Voyager 1　旅行者 1 号

CGRO: Compton Gamma-Ray Observatory　康普顿伽马射线天文台

CSD: CubeSat Design Specification　立方体卫星设计规范

CXO: Chandra X-Ray Observatory　钱德拉 X 射线天文台

GNSS: Global Navigation Satellite System　全球导航卫星系统

GPS: Global Positioning System　全球定位系统

GLONASS:　格洛纳斯（俄罗斯导航卫星系统）

SST: Spitzer Space Telescope　斯皮策空间望远镜

Section 6. Extended Reading Material

1. NASA's Spitzer Space Telescope Ends Mission of Astronomical Discovery

After more than 16 years studying the universe in infrared light, revealing new wonders in our solar system, our galaxy, and beyond, NASA's Spitzer Space Telescope's mission has come to an end.

Mission engineers confirmed at 2:30 p.m. PST (5:30 p.m. EST) Thursday the spacecraft was placed in safe mode, ceasing all science operations. After the decommissioning was confirmed, Spitzer Project Manager Joseph Hunt declared the mission had officially ended.

Launched in 2003, Spitzer was one of NASA's four Great Observatories, along with the Hubble Space Telescope, the Chandra X-ray Observatory and the Compton Gamma Ray Observatory. The Great Observatories program demonstrated the power of using different wavelengths of light to create a fuller picture of the universe.

"Spitzer has taught us about entirely new aspects of the cosmos and taken us many steps further in understanding how the universe works, addressing questions about our origins, and whether or not are we alone," said Thomas Zurbuchen, associate administrator of NASA's Science Mission Directorate in Washington. "This Great Observatory has also identified some important and new questions and tantalizing objects for further study, mapping a path for future investigations to follow. Its immense impact on science certainly will last well beyond the end of its mission."

Among its many scientific contributions, Spitzer studied comets and asteroids in our own solar system and found a previously unidentified ring around Saturn. It studied star and planet formation, the evolution of galaxies from the ancient universe to today, and the composition of interstellar dust. It also proved to be a powerful tool for detecting exoplanets and characterizing their atmospheres. Spitzer's best-known work may be detecting the seven Earth-size planets in the TRAPPIST-1 system – the largest number of terrestrial planets ever found orbiting a single star – and determining their masses and densities.

In 2016, following a review of operating astrophysics missions, NASA made a decision to close out the Spitzer mission in 2018 in anticipation of the launch of the James Webb Space Telescope, which also will observe the universe in infrared light. When Webb's launch was postponed, Spitzer was granted an extension to continue operations until this year. This gave Spitzer additional time to continue producing transformative science, including insights that will pave the way for Webb, which is scheduled to launch in 2021.

Vocabulary

infrared light：红外线

solar system：太阳系

Spitzer Space Telescope：斯皮策太空望远镜

safe mode：安全模式

decommissioning：退役

observatory：天文台

cosmos：宇宙

associate administrator：副局长

Science Mission Directorate：科学任务局

tantalizing：诱人的

comet：彗星

asteroid：小行星

unidentified ring：未确认的环

Saturn：土星

interstellar dust：星际尘埃

exoplanet：（太阳系外的）外部行星

terrestrial planets：类地行星

astrophysics：天体物理学

2. China's Lunar Rover Travels Over 284 Meters on Moon's Far side

China's lunar rover Yutu-2 has driven 284.66 meters on the far side of the moon to conduct scientific exploration on the virgin territory.

Both the lander and the rover of the Chang'e-4 probe switched to its dormant mode for the lunar night on Friday (Beijing time), according to the Lunar Exploration and Space Program Center of the China National Space Administration.

China's Chang'e-4 probe, launched on Dec. 8, 2018, made the first-ever soft landing on the Von Karman Crater in the South Pole-Aitken Basin on the far side of the moon on Jan. 3.

A lunar day equals 14 days on Earth, a lunar night the same length. The Chang'e-4 probe switches to dormant mode during the lunar night due to lack of solar power.

During the ninth lunar day of the probe on the moon, the scientific instruments on the lander and rover worked well, and a new batch of 2.9 GB scientific detection data was sent to the core research team for analysis.

As a result of the tidal locking effect, the moon's revolution cycle is the same as its rotation cycle, and the same side always faces Earth.

The far side of the moon has unique features, and scientists expect Chang'e-4 could bring breakthrough findings.

The scientific tasks of the Chang'e-4 mission include low-frequency radio astronomical observation, surveying the terrain and landforms, detecting the mineral composition and shallow lunar surface structure and measuring neutron radiation and neutral atoms.

The Chang'e-4 mission embodies China's hope to combine wisdom in space exploration with four payloads developed by the Netherlands, Germany, Sweden and Saudi Arabia.

Vocabulary

lunar rover：月球车

far side of the moon：月球背面

lander：着陆器

probe：探测器

dormant mode：休眠模式

lunar night：月夜

soft landing：软着陆

solar power：太阳能

tidal locking effect：潮汐锁定效应

revolution cycle：旋转周期

scientific task：科学任务

low-frequency radio astronomical observation：低频射电天文观测

neutron radiation：中子辐射

3. Europe and US Teaming up for Asteroid Deflection

Asteroid researchers and spacecraft engineers from the US, Europe and around the world will gather in Rome next week to discuss the latest progress in their common goal: an ambitious double-spacecraft mission to deflect an asteroid in space, to prove the technique as a viable method of planetary defence.

This combined mission is known as the Asteroid Impact Deflection Assessment, or AIDA for short. Its purpose is to deflect the orbit of the smaller body of the double Didymos asteroids between Earth and Mars through an impact by one spacecraft. Then a second spacecraft will survey the crash site and gather the maximum possible data on the effect of this collision.

The three-day International AIDA Workshop will take place on 11-13 September, 2019 in the historic surroundings of the 'Aula Ottagona' in central Rome, part of the Baths of Emperor Diocletian which went on to serve as a planetarium in the last century.

Participants will share the current progress of the two spacecraft making up AIDA - plus the smaller nano-spacecraft they will carry aboard them - as well the latest results of global astronomical campaigns undertaken to learn more about the distant Didymos asteroids.

NASA's contribution to AIDA, the Double Asteroid Impact Test, or

DART spacecraft, is already under construction for launch in summer 2021, to collide with its target at 6.6 km/s in September 2022. Flying along with DART will be an Italian-made miniature CubeSat called LICIACube (Light Italian CubeSat for Imaging of Asteroids) to record the moment of impact.

Then will come ESA's part of AIDA, a mission called Hera which will perform a close-up survey of the post-impact asteroid, acquiring measurements such as the asteroid's mass and detailed crater shape. Hera will also deploy a pair of CubeSats for close-up asteroid surveys and the very first radar probe of an asteroid.

Vocabulary

double-spacecraft：双航天器

deflect：转向

viable：可行的

planetary defence：行星防御

Asteroid Impact Deflection Assessment：小行星撞击偏转评估

Didymos asteroids：迪蒂莫斯小行星

collision：碰撞

workshop：研讨会

planetarium：天文馆

nano-spacecraft：微航天器

miniature：微型的

close-up survey：详细调查

post-impact asteroid：撞击后的小行星

crater：撞击坑

Hera: 赫拉（主神宙斯之妻）

4. Hayabusa-2 Completes Second Asteroid Touchdown to Collect Samples

Japan's asteroid-circling probe successfully executed a second touchdown on Thursday, June 3rd, 2019.collecting another sample from the surface of the space rock.

"The state of the spacecraft is normal and the touchdown sequence was performed as scheduled," the mission announced on Twitter. "Project Manager Tsuda has declared that the 2nd touchdown was a success!"

Japan's space agency, JAXA (Japan Aerospace Exploration Agency), shared images captured during touchdown on the mission's homepage. The photograph captured four seconds after touchdown shows debris from the asteroid's surface being flung into space.

The touchdown marked the second time the spacecraft has captured rocks and dust from the surface of Ryugu, a potentially hazardous asteroid in the Apollo group. The last touchdown and sample collection happened in February. Hayabusa-2 first rendezvoused with the asteroid in 2018.

The spacecraft is scheduled to leave its orbit around Ryugu in December and return to Earth in 2020. Though scientists are interested in more precisely plotting the asteroid's trajectory and the risk of its orbit intersecting Earth, researchers are most keen on studying the asteroid's composition.

"Studying Ryugu could tell humanity not only about Ryugu's surface and interior, but about what materials were available in the early Solar System for the development of life," according to NASA.

Ryugu samples could help scientists better understand how carbon-rich asteroids like it migrate from distant asteroid belts.

"We believe carbon-rich asteroids may have significant amounts of water locked up in their rocks. It's possible such asteroids may have brought to Earth both the water and the organic material necessary for life to start," Alan Fitzsimmons, astronomer at Queen's University Belfast, told BBC News. "These samples will be crucial in investigating this possibility."

Vocabulary

asteroid-circling probe：小行星环绕探测器

touchdown：着陆

space rock：太空岩石

touchdown sequence：着陆程序

debris：空间碎片

Ryagu：龙宫（位于地球和火星之间的一颗小行星）

rendezvous：交会

trajectory：轨道

composition：组成成分

carbon-rich asteroids：富含碳的小行星

migrate：迁移

organic material：有机物质

5. European Galileo Satellite Navigation System Resumes Initial Services

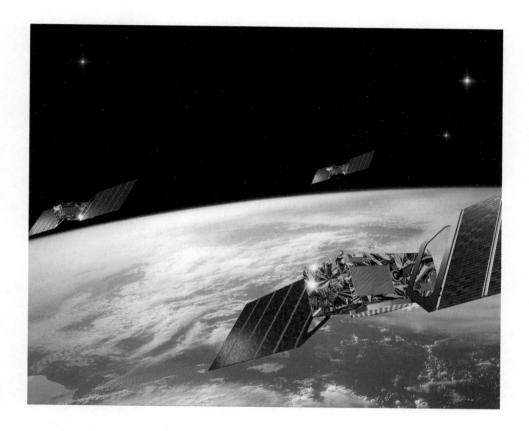

The Initial Services provided by the European satellite navigation system - Galileo - have been successfully restored. Galileo was affected by a technical incident related to its ground infrastructure. This event led to a temporary interruption of the globally available Galileo navigation and timing services, with the exception of the Galileo Search and Rescue Service.

The Search and Rescue Service, which is used to locate and assist people in emergency situations, for example, at sea or in remote, mountainous areas, was not affected and remained operational.

The navigation service impact was caused by a malfunction of some equipment in the Galileo control centers, which generate the system time and calculate orbit predictions; these data are used to produce the navigation messages. The disruption affected various elements at the control centers in Fucino (Italy) and at the DLR (German Aerospace Center) site in Oberpfaffenhofen.

A team of experts from the Galileo Service Operator, led by Spaceopal GmbH, worked as quickly as possible and in close cooperation with the European GNSS Agency (GSA), as well as the industrial ground infrastructure providers together with the European Space Agency (ESA) to rectify the malfunction.

Due to the high technical complexity of the system and the in-depth analysis of the fault dependencies, these efforts took several days before the resumption of Initial Services on 18 July 2018.

An independent board of inquiry is now investigating the exact circumstances and root causes that led to the failure. The investigation is being conducted by the EU Commission and the GSA - the authorities that manage the program - in order to continuously improve the system during its initial service phase.

Galileo has been offering its Initial Service since December 2016. During this initial 'pilot' phase, which precedes the 'full-operational services' phase, Galileo signals are being used also in combination with those from other satellite navigation systems, allowing testing and the detection of potential technical challenges during the commissioning of the system until full deployment and operational capability has been reached.

Vocabulary

Initial Services：初始服务

Galileo：伽利略

infrastructure：基础设施

interruption：中断

navigation and timing services：导航与授时服务

Search and Rescue Service：搜救服务

mountainous areas：山区

orbit prediction：轨道预测

Service Operator：服务运营商

rectify：修正

in-depth：深度地

board of inquiry：调查委员会

root cause：根本原因

pilot phase：实验阶段

full-operational services：全面服务

commissioning：试运行

阿波罗任务登月舱和月球车
Lunar Module and Lunar Car in Apollo Mission

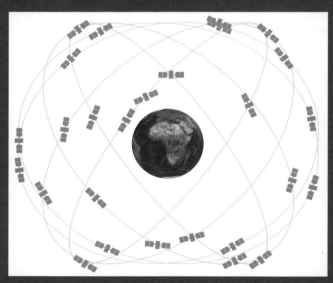

导航卫星系统星座
GNSS Constellation

好奇号火星车 Curiosity Mars Rover

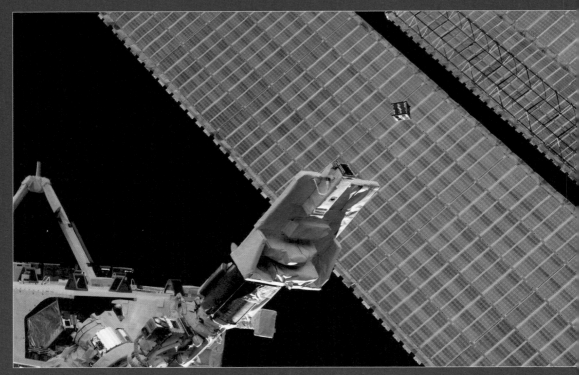

空间站释放立方星 Release Cubesat from ISS

太空中的月亮 Moon Seen in Space

维修哈勃太空望远镜 Servicing the Hubble
Space Telescope

意大利那不勒斯湾 Bay of Naples in Italy

银河系的中心 The Center of the Milky Way Galaxy　　　　　月球 Moon

追踪甲烷来源 Tracking Methane Sources

Chapter 6
Future Programs

太空探索永无止境，重返月球与登陆火星是载人航天的热点问题，美国提出并实施了"阿耳特弥斯"计划，瞄准2024年重返月球，已开展了航天发射系统、猎户座飞船以及"深空之门"月球轨道站的研制任务。载人登火对人类是一个巨大的挑战，保障航天员的健康与安全尤其困难，各航天大国已开展了相关技术研究，但2030年前很难实现。此外，寻找地外文明，也是人类的一个探索方向。

Section 1. Return to the Moon

场景介绍：重返月球已经成为世界载人航天的热点。凯文兴奋地向航天专家请教最新的登月计划——美国"阿尔特弥斯"任务的细节……

Kevin: Humans have landed on the Moon 50 years ago, why return to the Moon?

Ivan: Armstrong became the first human to step his feet on the Moon in 1969 during Apollo 11 mission. And humans have not been to the Moon ever since the last manned lunar mission-- Apollo 17 in 1972. Now, "return to the Moon" is becoming a hot spot again. The objective is not simply to land human on the Moon, but to verify related technologies and make preparations for the manned Mars mission.

Kevin: What programs have been planned for manned lunar mission?

Ivan: The U.S. government has declared that they will return to the moon in 2024 and NASA has proposed the Artemis program accordingly. The Russians say they will land cosmonauts on the moon but there is no concrete program. In China, advanced researches have been carried out, but there is also not a manned lunar program.

Kevin: So there is only one Artemis program, right?

Ivan: Yes. Artemis program is the only manned lunar program that is carried out.

Kevin: Can you explain the Artemis program?

Ivan: NASA's Artemis program plans to land the first woman and next man on the Moon by 2024. Then, they will use what they learn on and around the Moon to take the next giant leap – sending astronauts to Mars.

Kevin: What are the progresses of this program?

Ivan: They have recruited a new class of candidate astronauts. The launch facilities at Kennedy Space Center in Florida to support Artemis missions is upgrading, a powerful new rocket—the Space Launch System is designed and it will send humans and cargo to the Moon and beyond. The new Orion spacecraft is developed.

Kevin: What is the Gateway?

Ivan: Gateway is part of the Artemis program. It is also a vital part of NASA's deep space exploration plans, along with the Space Launch System (SLS) rocket, Orion spacecraft, and human landing system that will send astronauts to the Moon.

Kevin: Can you explain Gateway in more detail?

Ivan: The Gateway will weigh around 40 tonnes and will consist of a service module, a communications module, a connecting module, an airlock for spacewalks, a place for the astronauts to live and an operation station to command the Gateway's robotic arm or rovers on the Moon. Astronauts will be able to occupy it for up to 90 days at a time.

Kevin: So the astronauts will land on the Moon through the Gateway?

Ivan: Yes. Like a mountain refuge stock up with food and equipment for further climbs to the summit, the Gateway will provide shelter and a place to stock up on supplies for astronauts en route to more distant destinations.

Kevin: So the Gateway will not only support the manned lunar landing, but also other distant destinations, right?

Ivan: Yes. It may support the first manned Mars mission in the future.

Kevin: When will the first module of Gateway be launched?

Ivan: The Gateway's first module is set for launch on the second Artemis mission with Orion. The service module will fly on its own to the outpost's planned orbit and wait for the next module from the third Artemis mission.

Kevin: What about the progress of the Space Launch System and the Orion spaceship?

Ivan: The Artemis 1 mission will be launched in 2020 with the Orion spaceship on top of the Space Launch System. It will travel beyond the Moon and will stay for 3 weeks which will be the longest for a spaceship without docking to a space station module.

Kevin: What about the module development of the Gateway?

Ivan: NASA has focused Gateway development on the initial critical elements required to support the 2024 landing – the Power and Propulsion Element, the Habitation and Logistics Outpost (HALO) and logistics capabilities.

Kevin: Can you introduce the HALO? I think it is the crew cabin.

Ivan: Yes. The Habitation and Logistics Outpost will be the initial crew cabin for astronauts visiting the Gateway. It also will have several docking ports for visiting vehicles and future modules, as well as space for science and stowage. The HALO is being developed by Northrop Grumman and is managed by NASA's Johnson Space Center in Houston.

Kevin: What about the logistics to the Gateway?

Ivan: In March 2020, NASA announced SpaceX as the first U.S. commercial provider under the Gateway Logistics Services contract to deliver cargo and other supplies to the Gateway.

Kevin: What about the landing vehicle to the lunar surface?

Ivan: The Human Landing System is the final mode of transportation that will take astronauts to the lunar surface in the Artemis lunar exploration program.

Kevin: Where will the astronauts stay when they land on the Moon?

Ivan: On early missions, the astronauts will live inside the pressurized crew cabin portion of the lander for up to a week. In the future, the lunar habitat will be built.

Kevin: Who is developing the Human Landing System?

Ivan: NASA's Marshall Spaceflight Center in Huntsville, Alabama, was designated to lead NASA's Human Landing System (HLS) Program. On April 30, 2020, NASA announced selection of three companies to begin development on the Artemis Human Landing System: Blue Origin, Dynetics, and SpaceX.

Kevin: I have learned that the U.S. is inviting international cooperation in the Gateway program?

Ivan: Yes. Many countries or organizations, especially the current ISS (International Space Slation) partners, have decided to participate the Gateway program such as Canada, Japan and ESA (European Space Agency). They will provide important contributions to Gateway, comprising advanced external robotics, additional habitation and possibly other enhancements.

Kevin: What are the advantages of such a staging outpost near the Moon?

Ivan: Its flight path is a highly-elliptical orbit around the Moon – bringing it both relatively close to the Moon's surface but also far away making it easier to pick up astronauts and supplies from Earth – around a five-day trip. In addition, most current rockets do not have the power to reach the Moon in one go but could reach the space Gateway. They would be able to deliver supplies for astronauts to collect and use for further missions deeper into space – much like mountain expeditions.

Kevin: What spacesuits will the astronauts wear in Artemis mission?

Ivan: NASA's advanced exploration extravehicular mobility unity or xEMU will support lunar surface expeditions and during launch and re-entry, astronauts will wear the Orion crew survival system suit.

Kevin: Thank you for the introduction!

Ivan: You are welcome!

Section 2. Human Mars Landing

场景介绍：登陆火星是人类的梦想，但这也是对人类的巨大挑战。航天专家将解读载人探火的具体难点，并介绍目前已经开展的相关研究。

Kevin: What will be the next destination of deep space exploration for human being?

Ivan: The most promising planet is Mars.

Kevin: Why?

Ivan: Because Mars is the closest planet to the Earth and it is quite similar to our mother planet. Human have landed on the Moon and the Mars are considered by most of the space faring nations as the next manned exploration destination.

Kevin: How far is Mars to us?

Ivan: It is about 55-400 million kilometers away from the Earth because the two planets are both moving in different orbits around the sun.

Kevin: That is far more than the distance between the Earth and the Moon!

Ivan: Yes. If we travel to the moon, it will only take a few days and the communication is almost not affected, I mean there will be only about 1 second delay in communication since the distance is about 380000 kilometers. However, travel to Mars will take several months and the communication delay during Mars mission can be as much as 20 minutes. When there is an emergency, it is not possible to call the Earth for help, because they can only hear you 20 minutes later and when they call back it will take another 20 minutes.

Kevin: I have learned that the launch window to Mars is once in two years, is that right?

Ivan: Yes. Both the Earth and the Mars rotate around the sun with different orbital periods, and the distance between the Earth and Mars is constantly changing. We must take the shortest route. This will happen once in 26 months, approximately over years.

Kevin: When will the human being land on the Mars?

Ivan: It is hard to say. The Americans have declared that they will send human to Mars in 2030, but there is no concrete program.

Kevin: What are the challenges of a human Mars mission?

Ivan: The most challenging is for the human being.

Kevin: Can you explain in more detail?

Ivan: The trip to Mars and then coming back will take more than 500 days. That is extremely challenging psychologically for the astronauts to stay in a confined environment for such a long time. We must be sure that the crew can endure such long period time of travel.

Kevin: Can astronauts endure such a long trip?

Ivan: The answer is yes. In 2010, Russia, Europe and China have organized an experiment called Mars-500 in Mosco. A crew of 6 international volunteers participated a simulated flight to Mars including 240 days to Mars, 30 days on Mars and 250 days coming back. It proved that selected human with proper support can psychologically survive such long trip to Mars.

Kevin: That is the psychological challenge. What are the physical challenges to the body?

Ivan: The micro-gravity environment will cause a series of physical changes in the human body including bone loss and muscle atrophy. We know that for 6 months flight the physical changes of the human body are reversible thanks to the effective countermeasures such as exercise and medicine. That is why the crew rotates every 6 months on the ISS (International Space Station). However, the manned Mars mission will take one year and a half, so we must make sure that the crew can endure such long-term flights. To make things worse, the flight is not only in micro-gravity, there will be altered gravity, about 8 months microgravity, 1 month Mars gravity (one third Earth gravity), then another 8 months micro-gravity. That is rather challenging. Another big challenge is radiation.

Kevin: Why?

Ivan: In LEO (Low Eeath Orbit) orbit, the radiation is not a big issue, because the magnetic field of the Earth can shield most of the radiations, thus protect the astronauts in orbit. But once we left the protection of the Earth magnetic field on the way to Mars, the radiation becomes a big threat to the health of astronauts. The radiation encountered will be very complex including x-ray, γ-ray, proton, heavy ion etc. When there is a solar storm, it will be extremely dangerous. You know, large dose radiation may cause immediate death, and accumulated low dose of radiation may cause cancer and other diseases. The radiation may break the DNA in the human cells! The radiation protection for astronauts to Mars is not solved yet.

Kevin: That sounds horrible!

Ivan: The logistics is another problem for such a long-term mission. We must reduce the resupply as much as possible. That requires the development of Controlled Ecological Life Support System(CELSS) and in-situ resource utilization. CELSS means we can recycle the oxygen, water and food with an ecological system which could significantly reduce the logistic resupply for astronauts. In-situ resource utilization technology means we can use the local resources on the Mars to support the mission, for instance the water on Mars, the materials that can be used to produce the rocket fuels etc. That will help a lot for the logistics.

Kevin: What are the requirements for the astronauts?

Ivan: The requirement will be much higher than the LEO (Low Eeath Orbit) astronauts. The psychological requirements of Mars astronauts will be much higher, they should be able endure the lonely long flight, they will have to rely on themselves because of the communication delay. They must be able to endure the radiation, maybe there will be a gene selection to select astronauts that can better cope with radiation genetically.

Kevin: Thank you for the introduction!

Ivan: You are welcome!

Section 3. Looking for Extraterrestrial Life

　　场景介绍：寻找地外生命，是人类探索的一个新方向。凯文和航天专家探讨人类寻找地外生命的尝试，以及面临的争议和思考。

Ivan: Curiosity is the most important driven force of space exploration. "Are we alone in the universe?" Humanity has constantly asking the same question.

Kevin: Do we have the answer now?

Ivan: Not yet. The universe is unimaginably big and the astronomy distance is often measured by light-year, the distance that the light travels in one year! There are countless stars like our sun in the universe, and there are many earth-like planets in the habitable zone. We speculate that there must be other intelligent life in the universe but we have not had any evidence yet.

Kevin: How do we look for other civilizations in the universe?

Ivan: There are a number of large ground-based radio telescopes in the world and they are constantly listening to the signals coming from the universe.

Kevin: Can you introduce the radio telescope?

Ivan: The biggest one is the FAST radio telescope in Guizhou China, the full name of FAST is Five-hundred-meter Aperture Spherical radio Telescope. It is so sensitive that it can even see a match light on the moon 380000 kilometers away. Unfortunately, we have not received any signals of other civilization.

Kevin: Have we send our radio signals to the universe?

Ivan: Yes, signals have been sent many times, but there is no response. In addition, we have written information about human being on a disk including different pictures of human society and various sounds and put it inside Voyager 1 spacecraft, and hope that one day other civilization can get it and understand human being. Voyager 1 is now travelling out of the solar system and become the farthest man-made object.

Kevin: Some people believe that contact the aliens is dangerous, what is your opinion?

Ivan: Yes, many people worry that contact higher civilization may endanger the safety of humanity. Stephen Hawking, the famous physicist, has warned the danger of contacting the alien civilization. The famous science fiction *The Three Body Problem written* by Liu Cixin also express similar ideas. I think we must be cautious for the contact.

Kevin: One of the science objectives of Mars exploration is looking for possible lives. Are we looking for Martians?

Ivan: No. Looking for extraterrestrial life on Mars are more focused on microorganisms. There is a theory in astrobiology which speculates that life on Earth may originate from asteroids. Looking for life on Mars can help us to understand the origin of life.

Kevin: Can life survive the vacuum and extreme temperature environment in space?

Ivan: It is quite possible. Experiments on the ISS have showed that some life on Earth can survive the space environment for a long time such as tardigrade (water bear).

Kevin: If we explore the Mars, are there any risks of bring microorganisms of Earth to Mars?

Ivan: That is a very good question. If we bring the microorganisms from Earth to Mars in a space mission, the results may be disastrous. The contamination may destroy the ecosystem of Mars if there is life on Mars. So when humanity is sending landers or rovers to other planets, the spacecraft will be completely sterilized to prevent the possible biological contamination to the target planet. That is also true when we bring samples back from other terrestrial bodies. In Apollo missions, the returned crew and lunar rock samples are strictly isolated in a contained area to make thorough check up for fear of potential virus or microorganisms.

Kevin: Today, humanity mainly depends on chemical rockets to travel to space, but that is not possible for trips to other solar systems. Are there any other methods of space travel?

Ivan: Yes. Some new propulsion methods are proposed and technologies are becoming mature. While others are quite theoretical.

Kevin: Can you explain in more detail?

Ivan: The solar sail is a new method of propulsion. Although the acceleration is small but it is constant and the accumulated acceleration is satisfactory. Electrical thrusters and nuclear propulsion are also promising.

Kevin: What else?

Ivan: The scientists are also thinking that maybe we can travel by wormhole theory. That may solve the problem of traveling to place not possible for the current technology.

Kevin: Thank you for the explanation!

Ivan: You are welcome!

Section 4. Terminology and Abbreviations

Artemis program　阿耳特弥斯（月亮与狩猎女神）计划

Space Launch System（SLS）　太空发射系统

Gateway　深空之门（中途站、驿站）

human landing system (HLS)　载人着陆系统

Power and Propulsion Element　动力与推进模块

Habitation and Logistics Outpost (HALO)　居住与后勤前哨站

Exploration Extravehicular Mobility Unity (xEMU)　探索舱外机动装置

Mars-500　火星-500 实验

Extraterrestrial Life　地外生命（外星生命、天外生命）

Five-hundred-meter Aperture Spherical radio Telescope(FAST)

500 米口径球面射电望远镜

solar sail　太阳帆

FAST: Five-hundred-meter Aperture Spherical radio Telescope

500 米口径球面射电望远镜

HALO: Habitation and Logistics Outpost　居住与后勤前哨站

HLS: human landing system　载人着陆系统

SLS : Space Launch System　太空发射系统

xEMU: Exploration Extravehicular Mobility Unity　探索舱外机动装置

Section 5. Extended Reading Material

1. US spacecraft's solar sail successfully deploys

A new spacecraft's solar sail successfully deployed Tuesday one month after launch, as it journeys around earth carrying a small satellite.

LightSail 2, as the spacecraft is called, will harness the momentum of packets of light energy known as photons.

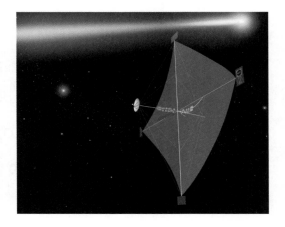

Its developer, the Planetary Society, a US organization that promotes space exploration, marked the occasion on Twitter with: "DEPLOYMENT COMPLETE!"

"All indications are that LightSail2 has successfully deployed its solar sail!" it said, based on data sent back to Earth from the spacecraft. The organization added that forthcoming photographs would confirm the event later in the day.

LightSail 2's solar sail has an area of 32 square meters and is made from Mylar, a brand of polyester that has been on the market since the 1950s.

Planetary Society Director Bill Nye, famous in American pop-culture for his eponymous children's show "Bill Nye the Science Guy," told AFP last month that the solar sail is "a romantic idea whose time has finally come."

"We hope this technology catches on." he said.

The solar sail will remain in orbit around Earth and, if all goes as planned, will gradually increase its altitude thanks to pressure from solar radiation on the sail.

As per its role in the future, Nye says the solar sail could one day be employed to send robotic missions beyond our solar system. While the initial speed of a spacecraft would be slower than if equipped with a motor, it would accelerate continuously and ultimately reach extraordinary speeds.

Vocabulary

solar sail: 太阳帆

harness: 利用

momentum: 动量

packet: 小包

photon: 光子

Planetary Society: 行星学会

indications: 显示

deploy: 展开

forthcoming: 即将来临的

Mylar: 密拉（一种聚酯薄膜）

polyester：聚酯

as per：按照

robotic mission：无人任务

initial speed：初始速度

accelerate：加速

extraordinary speed：非凡的速度

2. Sustaining Life on Long-Term Crewed Missions Will Require Planetary Resources

When astronauts live and work on the Moon, they will need access to life-sustaining oxygen, water and other resources. On the Moon, and eventually Mars, they could collect local resources on the surface and transform them into breathable air; water for drinking, hygiene, and farming; rocket propellants and more. It's a practice called in-situ resource utilization (ISRU).

In order to develop the best ways to find and use resources on the Moon, NASA Glenn Research Center in Cleveland created an Excavation Laboratory to explore effective ISRU methods using an Advanced Planetary Excavator robotic arm along with a shaker table and soil bins.

Using simulants of lunar soil, called regolith, researchers are testing excavation techniques and trying to understand how the digging process on Earth is different from the Moon or Mars.

"When we dig a hole into soil on Earth, we have plenty of weight, due to gravity, to put behind the shovel," explains Phil Abel, a team manager for the Excavation Lab. "On the Moon or Mars, we may not have the ability to put as much weight behind the shovel to dig effectively due to low gravity and the requirement to design light-weight machines for launch from Earth."

In addition to measuring the forces of excavation, researchers are also examining how to reduce those forces in order to reduce the necessary amount of vehicle traction.

Since astronauts will be living in space for longer and longer periods of time, carrying all the water, oxygen and rocket propellants needed to sustain life is not practical or affordable.

ISRU could reduce the mass and cost of launching products that are needed to sustain the crew. It could also help power outposts or refuel space vehicles. These critical supplies would be derived from space resources such as the carbon dioxide-rich Martian atmosphere and water on the Moon, Mars and asteroids.

As human space exploration evolves toward longer journeys farther from our home planet, ISRU will become increasingly important.

Vocabulary

life-sustaining：维持生命的

breathable：可呼吸的

propellant：推进剂

in-situ resource utilization：就地资源利用

Excavation：挖掘

Advanced Planetary Excavator robotic arm：先进的行星挖掘机机械臂

shaker table：振动台

soil bin：土壤箱

simulant：模拟物

regolith：风化层

shovel：铲子

light-weight：轻型

vehicle traction：车辆牵引

outposts：前哨站

refuel：补给燃料

carbon dioxide-rich：富含二氧化碳

3. Moon to Mars

The Gateway will enable months-long crew expeditions with multiple trips down to the lunar surface, enabling exploration of new locations across the Moon. The first part of the Gateway is targeted to launch on a private rocket in 2022 and will provide power and propulsion as well as communications for the spaceship. After it reaches orbit, and demonstrates its capabilities in space for about a year, NASA will launch astronauts in Orion on an SLS rocket carrying two new sections for the Gateway that will add a small living space and initial science and operational capabilities. Over time, the Gateway will become a way station for the development of refueling depots, servicing platforms, and a facility for processing samples from the Moon and other bodies in support of science and commerce.

SLS and Orion are critical to the NASA's exploration plans at the

Moon and beyond. NASA designed the Space Launch System as the world's most powerful rocket for safely sending humans on missions to deep space, and Orion is specifically designed to keep humans alive hundreds of thousands of miles from home, where getting back to Earth takes days rather than hours.

The first mission in 2020 will test the new spacecraft systems flying together for the first time, without crew, and the second flight is targeted for 2022 to take people for a flight test around the Moon. The third flight of SLS and Orion will kick-off delivery of new Gateway parts, with missions flying about once per year thereafter. SLS will launch the larger components for the Gateway on flights along with Orion, and Orion will be used as a tug to deliver those components to the required orbit for assembly. Together, Orion, SLS and the Gateway represent the core of NASA's sustainable infrastructure for human exploration.

NASA also continues to work with companies to address the challenges of living in space, such as using existing resources, options for disposing of trash, and more. Missions to the Moon are about 1,000 times farther from Earth than missions to the International Space Station (ISS), requiring systems that can reliably operate far from home, support the needs of human life, and still be light enough to launch. These technologies will become increasingly more important for the 34 million mile trip to Mars.

Exploration of the Moon and Mars is intertwined. The Moon provides an opportunity to test new tools, instruments and equipment that could be used on Mars, including human habitats, life support systems, and technologies and practices that could help us build self-sustaining outposts away from Earth. Living on the Gateway for months at a time will also allow researchers to understand how the human body responds in a true deep space environment before committing to the years-long journey to Mars.

Vocabulary

crew expeditions：飞行乘组探险

propulsion：推进

refueling depots：燃料补给仓库

servicing platforms：维护平台

Space Launch System：太空发射系统

tug：拖船

disposing：处理

intertwine：相互交织

human habitats：人居舱

life support systems：生命保障系统

4. NASA Eyes GPS at the Moon for Artemis Missions

GPS, a satellite-based navigation system used by an estimated four billion people worldwide to figure out where they are on Earth at any moment, could be used to pilot in and around lunar orbit during future Artemis missions.

A team at NASA is developing a special receiver that would be able to pick up location signals provided by the 24 to 32 operational Global Positioning System satellites, better known as GPS.

GPS is operated by the U.S. military about 12,550 miles above Earth's surface, and is open to anyone with a GPS receiver. These same GPS signals provide location data used in vehicle navigation systems, interactive maps, and tracking devices of all types, among many other applications.

Such a capability could soon also provide navigational solutions to astronauts and ground controllers operating the Orion spacecraft, the Gateway in orbit around the Moon, and lunar surface missions.

GPS is a system made up of three parts: satellites, ground stations, and receivers. The ground stations monitor the satellites, and a receiver, like those found in a phone or car, is constantly listening for a signal from those satellites. The receiver calculates its distance from four or more satellites to pinpoint a location.

Instead of navigating streets on Earth, a spacecraft equipped with an advanced GPS receiver may soon be paired with precise mapping data to help astronauts track their locations in the vast ocean of space between the shores of Earth and the Moon, or across the craterous lunar surface.

Navigation services near the Moon have historically been provided by NASA's communications networks. The GPS network, which has more satellites and can better accommodate additional users, could help ease the load on NASA's networks, thereby freeing up that bandwidth for other data transmission.

"What we're trying to do is use existing infrastructure for navigational purposes, instead of building new infrastructure around the Moon," said engineer and Principal Investigator Munther Hassouneh at Goddard Space Flight Center in Greenbelt, Maryland.

NASA has been working to extend GPS-based navigation to high altitudes, above the orbit of the GPS satellites, for more than a decade and now believes its use at the Moon, which is about 250,000 miles from Earth, can be done.

Vocabulary

satellite-based navigation system：基于卫星的导航系统

vehicle navigation systems：车辆导航系统

navigational solutions：导航解决方案

be paired with：与…配对

craterous：充满撞击坑的

Principal Investigator：首席研究员

5. Dragonfly Mission to Study Titan for Origins, Signs of Life

NASA has announced funding for the Dragonfly mission, featuring a drone-like rotorcraft lander that would explore the prebiotic chemistry and habitability of dozens of sites on Saturn's moon Titan.

The Dragonfly mission, part of NASA's New Frontiers program, will sample materials and determine surface composition to investigate Titan's organic chemistry and habitability, monitor atmospheric and surface conditions, image landforms to investigate geological processes, and perform seismic studies.

Planetary Science Institute Senior Scientists R. Aileen Yingst and Catherine Neish will be Co-Investigators on the Dragonfly mission.

Neish will study Titan's geology, with a particular focus on impact cratering, volcanism, and aqueous surface chemistry. Yingst will research what geologic processes have been - and currently are - active on Titan.

"Unlike other worlds we've landed on, Titan really has an otherworldly feel," Yingst said. "For a geologist, being able to study and remotely move around on the surface of a planet where water ice is as hard as rock, and liquid water would be considered a lava, is tremendously challenging and exciting."

"My Ph.D. dissertation investigated the creation of biological molecules on Titan's surface. Titan is a natural laboratory for the study of prebiotic molecules," Neish said. "I am thrilled to have the opportunity to 'collect the results' of these natural experiments as a part of the Dragonfly team."

Elizabeth Turtle, lead investigator on Dragonfly, worked at PSI (Planetary Science Institute) from 2002-2006 and is now at the Johns Hopkins Applied Physics Laboratory, which manages the mission for NASA.

The mission is slated to launch in 2026 and reach Titan in 2034.

Vocabulary

Dragonfly mission：蜻蜓项目

drone-like rotorcraft lander：类无人机旋转翼着陆器

Saturn：土星

Titan：土卫六（土星卫星之一）

surface composition：表面成分

geological processes：地质作用过程

seismic：地震的

volcanism：火山作用

aqueous：水的

otherworldly：超脱尘世的

lava：火山岩浆

molecules：分子

火星 500 试验密闭舱
Mars-500 Closed-cabin

火星 500 志愿者在舱内庆祝春节
Mars-500 Volunteers Celebrate Chinese
Spring Festival

猎户座飞船 Orion Spacecraft

火星探索 Mars Exploration

"火星一号"官网公布的人类未来在火星上的居住群区效果图
Renderings of the Future Human Habitats
on Mars Released by the Official Website of "Mars One"

火星 Mars